Beautiful Easy Lawns and Landscapes

Beautiful Easy Lawns and Landscapes

A YEAR-ROUND GUIDE TO A LOW-MAINTENANCE, ENVIRONMENTALLY SAFE YARD

by

LAURENCE SOMBKE

The Environmental Gardener

The Globe Pequot Press

Old Saybrook, Connecticut

Photographs by Catherine Herman
Cover and text design by Nancy Freeborn

Library of Congress Cataloging-in-Publication Data

Sombke, Laurence.
    Beautiful easy lawns and landscapes : a year-round guide to a low-maintenance, environmentally safe yard / by Laurence Sombke. — 1st ed.
        p. cm.
    ISBN 1-56440-357-2
    1. Lawns. 2. Landscape gardening. 3. Organic gardening. 4. Lawns—United States.
    5. Landscape gardening—United States. 6. Organic gardening—United States. I.Title.
    SB433.S62 1994
    635.9'64784—dc20   93-43687
                                                              CIP

Manufactured in the United States of America
First Edition/Second Printing

I want to dedicate this book to my wife, Catherine Herman,
who took the photographs and who helped me formulate so many of
the ideas that have been written about here.

Contents

Part One

ENVIRONMENTAL

LAWN CARE BASICS

Top Ten Ways
TO A
Beautiful Easy Lawn

Lawns just aren't what they used to be. That post–World War II, suburban, weed-free, closely cropped lawn-as-putting-green is a notion of the past. It is completely out of fashion from an aesthetic point of view, and it demands too much labor, time, fertilizer, water, and toxic pesticides to be practical in today's environmentally aware and busy world.

People today want lawns that are easy to take care of. Most people just barely have time to *cut* the darn grass, let alone water it, fertilize it, and manicure it. In addition to these practical considerations, there is genuine concern that toxic synthetic insecticides, herbicides, and fungicides can poison children, pets, and birds and other wildlife and pollute our drinking water. People are worried that synthetic chemical fertilizers will pollute rivers, lakes, bays, oceans, and other waterways. Many localities have the same concerns and have placed restrictions on the use of lawn chemicals in residential areas.

But people still want their lawns to look good. They don't want their neighbors to say, "Your lawn looks like hell, Bob." Because, after all, a crummy-looking lawn can make your neighborhood look shabby and actually lower the property value of the surrounding homes. Some neighborhoods even have regulations against bad-looking lawns.

That brings us to the key question: "How can I have a beautiful lawn that is easy to maintain without the use of chemicals?" You're in luck. This chapter has all, or at least most, of the answers you are looking for. Not only is it possible to have a beautiful, easy, chemical-free lawn, but it is also practical.

To try to make some sense out of how to get the job done, I have organized the information in this handy step-by-step program I call "Top Ten Ways to a Beautiful Easy Lawn."

TIP NUMBER ONE: *Stop Using Chemicals*

If you really want to have a beautiful lawn that is easy to maintain, you need to stop using chemical fertilizers, insecticides, herbicides, and fungicides. These applications may work in the short term, but in the long run your lawn will be a lot better off without them.

I know there's a battle raging in many families. One member says they must use chemicals or the lawn will look bad. Another family member is concerned about chemicals on the lawn and wants to stop using them. Chemicals have been around for a long time, but I have seen plenty of lawns that use them and still have weeds, diseases, and other problems.

Reliance on chemicals takes your lawn in the wrong direction. Because of the way chemicals work, many people tend to spray first and ask questions later. Thousands of pounds of chemicals are applied to lawns each year as part of a routine that people have adopted, not because there is an absolute need for them. In reality, the more you use chemicals on your lawn, the more your lawn becomes dependent on them. They become a kind of life-support system that seems to have no end in sight. There are better ways to make your lawn more beautiful, and the very first step is to switch your lawn over to a health-food diet and exercise program, just as many of us have done for ourselves in the 1980s and 1990s.

Your lawn has its own natural, healthy defense system that will come to the fore only if you stop using chemicals. The main problem with chemicals is that they tend to kill off the natural microorganisms, fungi, insects, and other "good guys" that can help your lawn fight off bad bugs, diseases, and other problems if only given the chance. If you drench your lawn with toxic chemicals, deprive it of organic matter, and water it too frequently, its natural systems will be disrupted; the turf will become weak and stressed out and become a breeding ground for bugs, diseases, and especially weeds.

Your lawn is a lot like your body. If you take care of yourself, get a proper amount of sleep, eat good foods with lots of fiber and vitamins, get some good exercise, and indulge in vices such as alcohol only in moderation, your body will be able to fight off most diseases and stress problems all by itself. But if you stay up all night, drink too much, smoke cigarettes, eat junk food, and generally abuse your health, that's when you are most likely to get diseases, even if it is only a common cold or the flu.

The same goes for your lawn. If you eliminate the use of chemicals and instead feed your lawn with natural organic fertilizer, add plenty of organic matter each year, grow the naturally disease- and pest-resistant grass seeds that are now readily available, and water correctly, you will be creating the kind of soil and grass-growing environment that has its own natural immune system. Your lawn will be able to fend off most pests and diseases all by itself.

Once you stop using chemicals and start your lawn on the way to recovery, your grass may suffer through a detoxification period while building up its soil health and microorganic environment. But then again it may not. In any case, this downtime, which I call conversion, should not last more than one growing season, and by the end of the year you should start seeing results. Remember that when you first started your exercise program, you didn't get results right away either, but pretty soon you started noticing changes in your body and your outlook. So stop abusing your lawn and instead put it on a health food and exercise plan that will pay big dividends for you and the environment in the very near future.

TIP NUMBER TWO: *Don't Treat Your Soil Like Dirt*

People tend to forget, or they just don't know, that their lawn is really a garden, too. Instead of having six tomato plants and a few peppers, a lawn has millions and millions of tiny grass plants. Each grass plant has to be treated with the same respect, tender loving care, and good old garden sense as those other plants.

In other words, you have to start the habit of improving and enriching the soil underneath the lawn where your grass plants grow. This is a simple two-step process: You need to make sure, first, that your soil's pH balance is perfect for growing grass, and, second, that your soil is rich enough in organic matter to provide the nurturing kind of environment that your grass plants will thrive in.

Check and Adjust Your Soil's pH Balance

The pH balance of your soil is a measure of its acid and alkaline levels. It is rated on a scale from 1 to 14, with neutral or "base" being pH 7.

A soil that is high in acid, also called a "sour" soil, will have a level below pH 7. A soil that is high in alkaline, also called a "sweet" soil, will have a pH level above 7. Most grass grows best in a soil with a pH level that ranges from 6.0 to 6.8.

It is uncommon for backyards in North America to have soils that are high in alkaline except in the Great Plains and Mountain states. It is very common for most of New England, the Northeast, the Middle Atlantic states, much of the the Middle West, and large parts of the South to have soils that are sour—that are high in acid and show a low pH level.

Having said that, there are vast differences in soil from state to state, county to county, and even backyard to backyard. The best way to find your soil's pH balance is to have the soil tested. A simple soil test is no more difficult or complicated than an early pregnancy test. You simply dig up several 1-pint measures of soil from different parts of your lawn, remove any grass or debris, let the soil dry out, and then place them in separate plastic bags. Mix the soil with the testing solutions and read the results according to the package instructions.

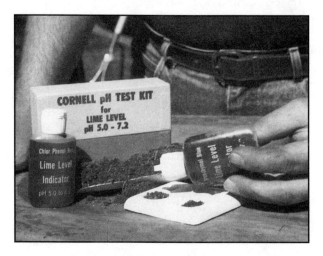

You should have your soil tested before planting a new lawn and then every other year for an established lawn to make sure the pH level is balanced. Simply dig up several cups of soil from various parts of your lawn, mix it together, and take it to your Cooperative Extension office. For a nominal fee they will test your soil and offer advice on correcting any imbalance. You could also use one of these handy soil test kits available at many lawn-and-garden centers.

If you want to have a soil test done for you, call your local lawn-and-garden center or Cooperative Extension office and ask if they are conducting soil tests. Many do, and they can cost as little as $1. A do-it-yourself kit can cost from $5 to $25. If you have the test done for you, a qualified expert can tell you how much lime, gypsum, or horticultural sulfur you need to add to correct any imbalance.

Many lawn problems are caused by a pH balance that is out of whack, which will mean that the nutrients your grass plants need to survive are chemically bound up in the soil and not available to them. This weakens the grass plants and gives weeds, which are not as sensitive to pH, the chance they are looking for to take over the yard.

Also, the millions of tiny microorganisms that digest the fertilizer and organic matter in your lawn thrive in soil with a pH balance of 6.0 to 6.8. Without these microorganisms your lawn will be neither beautiful nor easy.

If your soil test shows that your soil is below 6.0 on the pH scale, it means you should add lime to correct the imbalance. The best times to apply limestone to an established lawn are late summer, fall, and early spring. The best time to apply limestone to a new lawn is when you are adding the organic matter and fertilizer so that the limestone can be dug into the top 4 to 6 inches of soil.

Pelletized limestone is the easiest to use; you simply apply it with your drop spreader. Powdered limestone is also fine to use, although it is dusty and more messy.

For a new lawn the general rule of thumb is to apply 50 pounds of ground limestone per 1,000 square feet of lawn area to raise the pH level one point. For instance, if your soil pH tests at 5.5, you will need to add 50 pounds of limestone per 1,000 square feet of lawn area if you have a sandy loam soil, more if you have a mostly clay soil. Follow this table:

Pounds of ground limestone per 1,000 square feet of *new* lawn needed to raise soil pH to 6.5. Use half of these amounts to correct soil pH on an *existing* lawn.

SOIL pH	SANDY	LOAMY	CLAYEY
6.0	15	40	55
5.5	45	60	100
5.0	70	90	150
4.5	80	100	180

Once you've corrected your soil pH, you should remember to take a soil test every year or every other year and apply 10 pounds of pelletized ground limestone per 1,000 square feet every year or every other year to maintain your balanced pH level.

If your soil test shows that your soil is above 7.5 on the pH scale, it means you should add sulfur or gypsum to correct the imbalance. As with lime, the best times to apply sulfur or gypsum to an established lawn are late summer, fall, and early spring. The best time to apply sulfur or gypsum to a new lawn is when you are adding the organic matter and fertilizer so that the sulfur or gypsum can be dug into the top 4 to 6 inches of soil.

Again, pelletized sulfur or gypsum is the easiest to use. Apply it with your drop spreader. Powdered sulfur or gypsum is dusty and more messy, although acceptable to use.

For a new lawn the general rule of thumb is to apply 10 to 25 pounds of ground sulfur or gypsum per 1,000 square feet of lawn area to lower the pH level one point. So if your soil pH tests at 7.5, you will need to add 10 pounds of sulfur or gypsum per 1,000 square feet of lawn area if you have a sandy soil, 25 pounds if you have a mostly clay soil. This table will help you determine how much sulfur or gypsum you need:

Pounds of ground sulfur or gypsum per 1,000 square feet of *new* lawn needed to lower soil pH to 6.5. Use half of these amounts to correct soil pH on an *existing* lawn.

SOIL pH	SANDY	LOAMY	CLAYEY
7.5	10	15	20
8.0	25	35	45
8.5	36	45	55

Again, once you have corrected your soil pH, you should continue to take a soil test every year or every other year. To maintain the balanced pH level apply 10 pounds of pelletized ground sulfur or gypsum per 1,000 square feet every year or every other year.

Add Organic Matter to Your Lawn

It doesn't matter if you have a hard clay soil or a soft sandy soil. It is very important for you to add a yearly application of organic matter to transform your soil to sandy loam, which is a good balance of sand, clay, and organic matter. Loam soil is the very best for your lawn, and it is actually quite easy to come by. Your lawn likes to grow in sandy loam topsoil that is 4 to 6 inches deep and contains at least 5 percent organic matter. Most lawns grow in topsoil that is only 2 inches deep, is rather hard and compacted, and contains very little organic matter. Your lawn soil needs a lot of organic matter in it so that you can provide the right kind of environment for microorganisms to thrive, which will make your fertilizer more effective, which will attract worms that will naturally aerate your soil, allowing for deeper root penetration of grass plant roots, which will make your lawn more drought resistant. To get this you have to add organic matter every year. How?

First, don't rake up your grass clippings. Grass clippings are an excellent source of organic matter that will disappear and decompose in your lawn in a matter of days. If you cut your grass with a mulching lawn mower, they will disappear instantly and decompose in one or two days. Grass clippings are also a very good source of nitrogen fertilizer.

Next, in the autumn, when leaves are just beginning to fall, you can mow a few shredded leaves into the lawn. Leaves are another source of organic matter that will disappear in a matter of days and decompose in a matter of weeks. They are also a very good source of phosphorous, a nutrient that is important to the root growth of your grass.

Third, use natural organic fertilizer. Because it is made from dried meals, compost, and other organic agricultural byproducts, it is high in organic matter. Whenever you use natural organic fertilizer, you are adding organic matter back to the soil.

Finally, most lawns benefit from a yearly top dressing of organic matter. By top dressing, I mean ½ inch of sifted compost, composted humus, or dehydrated cow or horse manure spread over your entire lawn. Many people like to top-dress their lawns with peat moss, which is clean, easy to use, and fairly inexpensive, and it does a good job of helping to loosen the soil. Mixing compost with peat moss might be a good way to stretch your lawn-care dollar.

Top-dressing in spring or fall is a good way to add organic matter to your lawn to eliminate thatch buildup and to make the lawn more drought resistant. Simply rake a ¼- to ½-inch-thick layer of compost, composted manure, or dehydrated manure over your lawn. The top dressing will disappear in a matter of days as it settles into the ground and eventually the grass roots, making your lawn healthier.

Top dressing is simply a matter of opening several bags of organic matter, dumping the contents out over the lawn, and then raking it around so that the lawn is evenly covered with ¼ to ½ inch of material. You can also easily apply top dressing with a fertilizer spreader if you have one. Fall is the very best time to top-dress your lawn, but early spring is a good time, too.

TIP NUMBER THREE: *Give Your Lawn a Breath of Fresh Air*

Every good healthy lawn needs water, fertilizer, and air to be really beautiful. For most lawns, ample air is the missing element. The roots of grass plants need air to grow properly, and soil microorganisms need air to keep alive while digesting organic matter. The biggest problem for lawns is when the air is forced out of the soil by compaction, that is, smashing the soil down by running all over it the way kids love to do in the summertime. Another problem is the accumulation of thatch, that thick layer of dead roots that accumulates in a lawn that is overfertilized, overwatered, and spread with too many chemicals.

If your lawn feels soft and cushiony to walk on and you can't see the soil beneath the grass, you probably have a thatch problem. When you plant your annual flower or vegetable garden, you dig up the soil with a spade or rotary tiller, which breaks up the soil and adds plenty of air, but that is not practical for lawns, of course. Here are a couple of solutions:

Do Some Spring Cleaning on Your Lawn

Each spring, just as the weather begins to warm up and the grass begins to grow again, take a long-handled lawn rake to your yard and give it a good scratching. If your lawn is like mine, you will find whole leaves, twigs, rocks, bottle caps, and other assorted debris left over from summer and winter that can easily smother any grass that is growing again and create bare spots, which will soon be filled with weeds. Wear a pair of leather gloves to protect your tender hands, and start raking. Separate the bottle caps and rocks and place all the organic matter in your compost bin.

Aerate Your Lawn with a Tool

There are several tools you can use to aerate your lawn. For larger lawns, you can rent an aeration machine that digs tiny cores of dirt out of the lawn, allowing air, water, and nutrients to find their way deeper into the soil.

For smaller lawns, you might try using a hand-held core lawn aerator, which is something like a pogo stick with spikes on the end. You simply line the spikes up on the lawn, step on the foot pads just like on the back of a shovel, and force the spikes down into the soil. Pull it up and pop out the tiny bits of soil. A pitchfork and a pair of golf shoes are two of my favorite low-cost, low-tech ways of aerating the lawn. Garden expert Jerry Baker is an advocate of golf shoes. You simply wear your pointy spiked golf shoes out for a leisurely walk on your lawn. The pointy spikes poke little holes in the lawn, allowing air to penetrate. Or you

Aerate your lawn every year to break up a hard, compacted soil surface and promote better water retention and fertilizer usage. Rent a power aerator or simply poke holes in the ground with a forked spade. You golfers can aerate your lawn by walking on it with your spiked golf shoes.

can poke a pitchfork down into the soil and keep on poking to accomplish the same result.

It doesn't matter what tool you use; if your soil is compacted, you ought to aerate it to relieve the situation.

TIP NUMBER FOUR: *Fertilize the Natural Way*

Up until recently, most people fed their lawns with synthetic chemical fertilizers bought at their lawn-and-home centers or with manure hauled from the local farmer's barn. You could buy natural organic fertilizers in selected garden centers or from specialty garden catalogues, but they weren't widely available and their use was a little difficult to understand.

Now all of that has changed. Environmentalists say that synthetic chemical fertilizers can be a source of water pollution and that fertilizers premixed with pesticides can pose health and safety problems for people, pets, and wildlife.

Nowadays, most lawn-and-garden centers are well stocked with new lines of organic fertilizers for both lawns and gardens. They are made from recycled agricultural by-products such as bone meal, composted chicken manure, wheat germ, chicken feathers, and grain meals. They carry names such as Ringer Lawn Restore, Espoma, Nature's Best, Milorganite, Nature Safe, Nurture, EarthGro, and 1881 Select. Even Scott's, the venerable chemical fertilizer company, has a line of organic-based fertilizers called Next Generation. They are packaged in small 5-pound boxes for small gardens and in the familiar 20- to 25-pound bags for lawns and larger gardens.

Just like their chemical competitors, the natural organic fertilizers are granular, clean, and easy to use, and they don't give off offensive odors like barnyard manure does. You apply them just as you would chemical fertilizer, broadcasting them with a drop spreader or rotary spreader or sprinkling them on by hand. They also have the same coverage area as chemicals. For instance, a 25-pound bag of natural organic lawn fertilizer will feed 2,500 square feet of turf grass.

These new natural organic fertilizers are the best solution yet for a gardener or lawn keeper who is looking for an easy-to-use, nonchemical fertilizer that really gets high-quality nutrients to the grass. Plus, they have the added attraction of helping to solve our landfill crisis and keep our waterways clean. Here's how:

- Natural organic fertilizers are made from recycled agricultural by-products that would otherwise be dumped in our already overcrowded landfills. For instance, EarthGro in Connecticut gathers apple and cranberry pulp from the juice industry, manure and feathers from the poultry industry, leaves and sawdust, and other discards to make its fertilizer.

- Synthetic chemical fertilizers are made primarily from natural gas, a nonrenewable fossil fuel. The manufacture of fertilizer does not put a dent in the world's supply of natural gas, but some people are concerned about reducing their use of fossil fuels.

- Natural fertilizers provide more nutrients to the soil than animal manures by themselves do. For instance, blood meal, which is a primary component of Nature Safe's lawn fertilizer, has a much higher nitrogen content than cow or horse manure.

- The naturals release their nutrients to the plants slowly and over a longer period of time. Chemical fertilizers act more quickly but then are gone more quickly.

- Natural fertilizers won't burn your lawn or your plants with too much nitrogen. If you spill natural fertilizer on your lawn, it won't kill the grass.

- Natural fertilizers add much-needed organic matter back to the soil. This is more critical for lawns and perennial beds, because you can't just dump leaves and compost over the area and rototill it in every year to add organic matter like you can with a vegetable or annual flower garden.

- Natural fertilizers are more water conserving. First, they are not water soluble; they don't need to be watered in to work like many chemical fertilizers do. Second, the organic matter they return to the soil, especially in lawns, helps make the soil more like a "sponge," which can hold water better for use during drought.

- Naturals do not contribute to water pollution. Environmentalists estimate that as much as 50 percent of the nitrogen in chemical fertilizers never reaches the plants. Instead, it gets washed off into waterways, where it encourages the growth of aquatic plants, which use up the oxygen in the water, making it unsuitable for fish. Sewage is the main culprit of this eutrophication process, but chemical fertilizers are also being targeted as a potential source of pollution around Long Island Sound and Chesapeake Bay, for example.

The biggest criticism of natural organic fertilizers is that they appear to cost more than chemicals. It's true that they're priced higher. Chemical fertilizer usually runs less than $10 per 25-pound bag; naturals cost from $10 to $15, even as high as $20, per 25-pound bag in some stores. But the price differences may be misleading. Naturals feed the garden, especially lawns, over a longer period of time, meaning fewer applications are required. Naturals add organic matter to the soil, making it more water conserving, and they don't require watering in, which means you will save money on your water bill. With naturals, there is no chance of water pollution. When you factor in all these costs and savings, natural organic fertilizers are as good a buy as chemicals.

How do you know you're getting a natural organic fertilizer? Read the label. Look on the ingredients list for things like bone meal, fish meal, feather meal, soybean meal, sunflower seed-hull ash, and langbeinite. Chemical fertilizers will list urea, ammonium sulfate nitrate, and methylenedurea as ingredients. Take note of bags marked "natural-based" fertilizer. Scott's makes lawn fertilizer that is made primarily of natural organic products, but a small amount of chemicals has been added to the mix to achieve certain results. Scott's Next Generation flower and vegetable fertilizers are all natural. Buy what you want, but be sure to get what you pay for.

Natural organic fertilizer is made from recycled agricultural by-products. It feeds your lawn slowly, it doesn't burn the grass, and it makes your lawn lush and green.

If you can't find natural organic fertilizer at your local store, here are names and numbers of the major manufacturers:

Ringer, 9959 Valley View Road, Eden Prairie, MN 55344, (800) 654–1047.

Nature Safe, Cold Spring, KY 41076, (800) 252–4727.

EarthGro and 1881 Select, P.O. Box 143, Lebanon, CT 06249, (800) 736–7645.

Koos, Inc., Kenosha, WI 53141, (800) 558–5667.

O.M. Scott, 14111 Scottslawn Road, Marysville, OH 43041, (800) 874–7336.

In addition to garden-and-home centers and hardware stores, there are two good catalogue sources for natural organic fertilizer:

Gardens Alive! 5100 Schenley Place, Lawrenceburg, IN 47025, (812) 537–8650.

Gardener's Supply, 128 Intervale Road, Burlington, VT 05401, (802) 863–1700.

TIP NUMBER FIVE: *Let Your Grass Grow Taller*

This is the one tip that will cost you absolutely nothing and will pay you and your lawn enormous benefits. By simply letting your grass grow taller, up to the 3-inch level in many cases, you will reduce weeds and pests in your lawn, need to use less water and fertilizer, and have a lawn that is much more beautiful to behold.

How can something this simple be so valuable? Easy. Just remember that your lawn is made up of millions of individual grass plants. Each grass plant has leaf growth on top of the ground, that green stuff you have to cut so often, and root growth below the ground. You want grass plants that have deep root growth, which will translate into stronger, healthier grass that can withstand most diseases, pests, and drought.

Taller grass above the ground helps promote deeper root growth. The deeper the roots go, the more broken up and aerated the soil becomes, making it a better place for microorganisms to thrive, allowing more water to seep down into the soil, making your lawn more like a reservoir than a brickbat. What moisture you do get, either from the heavens or from the sprinkler, will stay in the soil longer and help the grass grow better.

For cool-season grasses, including Kentucky blue, perennial rye, fine fescue, and turf-type tall fescue, you can set your mower blades to cut the grass at 3 inches tall.

Set your mower blades to cut the grass at 2 to $2^1/2$ inches tall for Bermuda and zoysia grass and 3 inches for Bahia and St. Augustine.

In the late fall, when you are about to cut your grass for the last time of the season, lower your blades to 2 inches for cool-season grasses and 1 to $1^1/2$ inches for warm-season grasses. A shorter mowing in the fall prevents the grass from matting down over the winter, which could lead to disease in the spring.

Recycle those grass clippings by leaving them on the lawn. Don't bag them and drag them to the dump. Americans send more grass clippings to our landfills than either newspapers or plastic bottles. Grass clippings alone can give you 25 to 50 percent of your fertilizer needs absolutely free.

Grass clippings do not cause thatch. An overmanaged, chemically intensive lawn-care program causes thatch. Thatch is a natural $1/2$-inch zone where the grass meets the soil. It is where the microorganisms live. Excess thatch is made up of dead and damaged roots that are growing too near the surface. Tall grass promotes deep roots, helping to prevent thatch.

Tall grass creates the type of environment in which the grass clippings will decompose in just a matter of days. Your lawn becomes a mulching compost bin that is constantly decomposing organic matter and adding it back to the soil.

Always follow the one-third rule, that is, never cut off more than one-third of your grass's height at any one time. If you want your grass to be 3 inches tall, cut it at or before it reaches 4 inches. Taller grass can handle these short clippings and make them disappear in a matter of days if you have a simple side-discharge or push-reel mower. If you cut your grass with a mulching mower, the clippings will disappear instantly.

If you follow the one-third rule, it means you will have to cut your grass every four or five days rather than once a week during the surge of grass cutting in the spring. But research has found that you will spend 40 percent less time cutting your grass this way than all the time you spend bagging and dragging those clippings off to the landfill.

If you get behind on your grass mowing and can't follow the one-third rule, rake up the grass or reattach the bag and place those clippings in the compost bin. Large clumps of grass clippings left on the lawn can damage the remaining turf.

Keep your blades sharpened, maybe twice a year. Dull blades mangle the grass and can cause damage.

TIP NUMBER SIX: *Plant the Right Grass Seed*

I would guess that at least half of the problems you are having with your lawn are caused by growing the wrong type of grass in the wrong place. Chances are you didn't plant the seeds that produced the grass that now controls your lawn. You're probably just coping with the lawn that was already there, a lawn that might never have been planted, that could have been just cow pasture, an old orchard, or a cornfield that sent up some type of grass that is now your challenge.

Growing the right kind of grass can solve many of your problems, because there have been major advances in grass seed research over the past few years. Today's grass seed is more drought resistant, more disease resistant, and, better yet, a lot of it is even resistant to such enemies as chinch bugs, billbugs, sod webworms, and maybe even grubs.

Let's start at the beginning by finding out what type of grass you have. Easy. Simply dig up a slice of turf, wrap it in plastic, and take it to your Cooperative Extension office or favorite lawn-and-garden center. Chances are you've got the most common grass plants in your area, and an expert will have no trouble telling you what you have. Once you have determined what's out there, ask yourself these questions:

Where do I live? Cool-season grasses like Kentucky blue grow better in the northern parts of the United States and Canada. Warm-season grasses like

Bermuda grow better in the deep South. But there are a lot of transition zones between north and south where many types of grasses grow well.

Is my yard mostly shady or mostly sunny? When your lawn was created, as many as thirty or fifty years ago, there might have been no big trees in your neighborhood. Now those little twigs have become mighty oaks, and you might have sun-loving grass growing in a shady yard. Certain grasses do better in sun, and others do better in shade.

What sort of wear and tear does my lawn get? If you've got kids or grandkids who want to play baseball or soccer out there, you need to grow grass that can take some abuse.

With these answers in mind, you can start making some decisions. It doesn't matter if you are going to salvage your existing lawn by overseeding it, or dig up the north forty and plant a whole new lawn: You still need good grass seed. Here are guidelines that should help you make a decision.

Good Grass Guideline Number One: Buy Expensive Grass Seed

With grass seed, you get what you pay for. Better-quality grass seed costs more money. Look for brand names that you can trust or that your seed dealer will stand behind and buy the best seed in the line. Cheap grass seed will have a poorer germination rate, and it will contain a higher proportion of chaff to seed in the package. You are going to spend a lot of money on equipment, organic matter, lime, fertilizer, and water to plant your lawn, plus untold hours of your time to do the work, it doesn't make any sense to skimp on grass seed. This same advice goes for sod, sprigs, or plugs, if you are going to use them instead of seed to plant a new lawn.

Good Grass Guideline Number Two: Buy a Blended Mixture of Different Types of Grass Seed

Except for warm-season grasses, which do not like to be blended, you want a blend of different types of grass seed, so that if one type gets attacked by a pest or a disease, the others will live on and thrive. Look at the label on the side of the seed box. In cool-season grass areas, look for a mixture of perennial ryegrass, fescue, and Kentucky blue—maybe one or two types of each of these varieties of grass seed. Warm-season lawns have fewer varieties of grass seeds to choose from, and, in general, warm-season grasses do not like to be grown in a blend.

In addition, you want to look for named grass seed, such as Titan or Adelphi, because these are the improved varieties. If the grass seed has a name, it usually is a better-quality product.

Good Grass Guideline Number Three: Look for Seeds Treated with Endophytes

Endophytes are naturally occurring fungi that have been found to repel most chinch bugs, billbugs, fall army worms, sod webworms, and aphids. Researchers are investigating whether or not endophytically treated grass seed can repel grubs. One again you have to read the label or ask your most knowledgeable lawn-and-garden center professional. Some seed companies are trumpeting their treated grass seed, and some are being more sedate about their products. As yet, fine fescue, perennial rye, and turf-type tall fescue are the only types of grass seed that are treated with endophytes, but I imagine that will change if scientifically possible.

In addition, you should look for grass seeds that are disease resistant and drought resistant. Once again, the label will tell you which is which.

Good Grass Guideline Number Four: Plant at the Right Time

You can't just plant grass seed for a new lawn any time you want. For the best chance for success, you should plant cool-season grasses like fescue, Kentucky blue, and perennial rye in the late summer or early fall, six weeks before the first expected frost. This will give the grass time to get established before the cooler weather sets in. Springtime is the second best time to plant cool-season grasses, although it is risky because spring is when annual weeds such as crabgrass germinate and would love to take over your lawn. It is best to plant warm-season grasses like Bermuda, zoysia, and St. Augustine in spring and early summer, between late March and early June. That gives them a good chance to get established during the height of their growing season.

TIP NUMBER SEVEN: *Water with Conservation in Mind*

We've all done this: You come home from work on one of those "dog days" of July, it's hot, the grass looks wilted and in need of some water. You hop out of the car, grab the hose, and spray the lawn for ten minutes, until you get bored and the lawn looks a little moist. What you're really doing is wasting water and ruining your lawn.

First of all, you're watering at the wrong time of day. If it is sunny and windy, you may have lost half of that water to evaporation.

Then, you're watering the green tops of the grass and not the roots, and that doesn't do the lawn any good at all.

But worst, you're coddling your lawn and turning it into a wimp. The roots of your grass plants will go wherever they have to to get water. If you make it easy on them and give it to them right up close to the surface, that's where they will grow. But they won't grow strong. They will get lazy, they will get hooked on easy water, and the first time there is a drought, they will wimp out on you and cause your lawn to deteriorate into stress. A stressed lawn is an open invitation to bugs and diseases.

Instead of being an easy water pusher to a wimpy lawn, you need to toughen up your grass and force the roots to dig down deep in the soil to find that water. When the roots grow deep, they get stronger and more able to fend for themselves during drought. But they also break up the soil, creating a water reservoir underground that the plants can call on when rain is scarce.

Here's the secret: Water the roots not the grass. You do that by delivering water as close to the root zone as possible, by watering as slowly as possible and as infrequently as possible.

Soil will absorb water only at a certain rate. It's just science. A clay soil absorbs water very slowly but will hold more than a sandy soil, which sheds water very rapidly. What you want to avoid is runoff. How many times have you driven down the street and seen water running down the gutter from a sprinkler that is watering somebody's driveway or sidewalk? That's crazy! Water is too precious to waste, and it doesn't do your lawn any good at all.

In general, your lawn needs to consume 1 inch of water each week. If you get rainfall, that might just be enough. If you don't, you might need to supplement.

But first, a word to the wise. If you have a generally healthy lawn, and you will if you follow my ten-point program, you can allow your cool-season grass lawn, that means most of you from Maine to Macon and west to the Rockies, to go dormant during July and August. That's right. It is natural for your lawn to turn light brown during the driest summer months, only to return to luscious green when the rain resumes in September. My advice: Don't water your lawn during the summer. Especially if you have well water. Watering your lawn is not worth having your well go dry. And if you think about it, all of us are hooked up to a big well or a reservoir or some other water source that could also go dry during drought.

If you insist on watering your lawn, here's the best way:

To deliver that water as close to the root zone as possible, you need the sprinkler best equipped to do that. If you can afford it, a built-in ground irrigation system is the best. Especially the ones that mist the water in a light spray or drip or have pop-up impulse sprinklers. An in-ground irrigation system is very water efficient, and it adds to the value of your property.

The next best tool is an impulse sprinkler, the kind that is used on golf courses that shoots the water out in a big circle and then resets itself. An impulse sprinkler is good because it has a low-to-the-ground trajectory of spray, and it gives a good, even distribution of water. Buy one that is made out of solid brass, because it will last much longer than one made out of breakable plastic.

Oscillating sprinklers are easily the most common but not necessarily the best. They tend to be difficult to set for the right coverage area—they water the garage door, driveway, and sidewalk, for example—and they tend to water too much at the top of the arc and not enough at the end of the arc. Some of the newer and more expensive oscillating sprinklers, such as those by Gardena, do a much better job delivering water than their cheaper predecessors.

Do not water your lawn with one of those twirling sprinklers. They are fine for the kids to play in, but they just don't give an even distribution of water over a very large area.

The best time to water is in the early morning to avoid the sun's evaporating rays and to give the water time to soak in before the heat of the day causes too much evaporation. It's okay to water in the evening or at night as long as your lawn is not prone to disease problems.

Water no more frequently than once a week, and let the sprinkler run for one to two hours or however long it takes to soak the ground to a depth of 6 to 8 inches. (In general, 1 inch of moisture applied at the correct rate will soak the ground to a depth of 6 to 8 inches.) You can gauge this by poking a little hole in the ground and measuring for moisture depth, or you can set out a cake pan in the watering area and stop watering when the pan has collected 1 inch of water.

TIP NUMBER EIGHT: *Nip Weeds in the Bud*

The only way to prevent the spread of weeds in a beautiful easy lawn is to stop them before they get started. Weeds are opportunists. All lawns contain millions of tiny weed seeds, lurking in the soil, just waiting for their day to come. As soon as your grass becomes weak from too much of the wrong kind of fertilizer, too little organic matter, a soil pH that is out of balance, superficial watering, or, worst of all, low mowing, weeds are more than happy to step in and take over. The real key to controlling weeds without the use of chemicals is to make your lawn so vigorous, healthy, and strong that your grass emerges victorious in the battle for a beautiful easy lawn.

For instance, most crabgrass can be prevented by simply mowing the grass at a 3-inch height rather than 1½ to 2 inches. Why? Because crabgrass is an annual weed, which means it has to germinate from a seed every year. To germinate, it needs sunlight. If your grass is 3 inches tall, it will create a dense shade that will make it impossible for crabgrass to sprout and grow. You don't have to spend good money on the thousands of pounds of chemical crabgrass preventer that are dumped on our lawns each year. You simply need to let your grass grow taller.

The idea of letting your grass grow taller to prevent crabgrass is just an example of the kind of thinking that is going on at university research centers across the United States. Scientists are looking for ways to reduce or eliminate the use of toxic chemical pesticides, and they are discovering that there are effective ways to control weeds through mowing, fertilizing, and watering changes.

Before I get into the specifics of how to control some of the most well-known weeds, let's take time out for a little philosophy. The concept of a weed-free lawn only became possible after World War II with the widespread introduction of synthetic chemical herbicides. Before that, people had weeds in their lawns, and that was all there was to it.

The ideal of a lawn without weeds is outmoded and behind the times. A pure, 100 percent all-grass lawn looks unnatural to me. I like a lawn to have a more natural look. Today's lawn contains a few weeds that I think add to the lawn's beauty and health. For instance, I have quite a few violets in my lawn and I love my violets. I look forward to their springtime beauty and I would be sad if they disappeared from my lawn. I also have ajuga, moss, and forget-me-nots in certain places. I draw the line, however, at plantain, which I consider to be unsightly, and I dig it out every chance I get.

Having said all that, here are some common lawn weeds (at least they're the ones that people most often ask me about) and how you can control them. Of course, the most important thing you can do to stop weeds is to adopt the seven tips listed earlier in this chapter: Stop using chemicals, build up a rich organic soil, aerate the lawn if necessary, use natural organic fertilizer, let your grass grow taller, grow the new stronger grass seeds, and water slowly and infrequently.

Crabgrass. This is by far the most common weed found in American lawns. It is also the easiest to cure without chemicals. Let your grass grow 3 inches tall. Tall grass prevents crabgrass seeds from getting the sunlight they need to germinate. Fertilize during early spring or late fall, never during summer when crabgrass is growing like gangbusters and turf grass is going dormant. Overseed or renovate your lawn in fall when crabgrass is dying, never in spring when crabgrass is primed for germination. Avoid superficial watering, which favors crabgrass growth.

Dandelion. A lot of people think dandelion has a pretty yellow blossom. Many people, particularly old-time Italian families, love to eat dandelion leaves in salads. Dandelion's deep taproot also aerates the soil and helps bring nutrients toward the soil surface, which can benefit turf grass.

Even with all its attributes, however, people still feel they need to reduce the infiltration of their lawns by the dandelion. To do that, dig dandelions up by hand when they are blooming, because that is when they are at their weakest. Keep digging at them often. Even if you don't get all that root the first time, constant digging will eventually deplete the plant's food supply and force it to die. Once the plant is out, sprinkle the area with new grass seed to grow and fill in the spot.

Plantain. Plantain doesn't have all the blessings that dandelion does, and you should dig it out as soon as possible.

White Clover. Leave it. A small percentage of white clover in your lawn, say 5 to 10 percent, is actually very beneficial. Clover helps bring much-needed nitrogen into the soil. It is very drought and stress tolerant, and it helps aerate the soil with its strong root system.

Wild Onion. Leave it or pull it up by hand, making sure you get the tiny white bulbs. You may want to dig wild onion after a good rain when the soil is soft. As a cook, I like to smell the fragrant scent of wild onion when I cut it with my lawn mower. Remember, the real secret to weed control is to keep that turf grass growing strong and tall. If more than 50 percent of your lawn is given over to weeds, you may want to completely remodel it, which you can read all about in the "Lawn Renovation, Restoration, Remodeling, and Rehabilitation" chapter. If you really feel you need to spray some weeds rather than pull them, try using Safer's insecticidal soap as an environmentally friendly alternative.

TIP NUMBER NINE: *Let the Good Guys Take Care of the Bad Guys*

Insects and diseases have the same nasty attitude and unpleasant disposition as weeds. They need weakened turf grass, depleted soil, and poor growing conditions to attack and get the upper hand on your lawn.

Nowhere is a healthy, vigorous, growing lawn more necessary than in the scheme to control bugs and diseases. But before you do anything, remember this: A few grubs and beetles will not ruin your lawn. Don't panic if you see a few dead blades of grass. The worst thing you can do is spray first and ask questions later. Here's what you can do:

First, maintain a healthy lawn by building the soil with organic matter, feeding it with natural organic fertilizer, watering slowly and deeply, and letting your grass grow taller.

Second, "Build it and they will come." By this I mean that insects and diseases have their own natural predators, like parasitic wasps, birds, and microorganisms. Synthetic chemicals create a lawn environment that does not welcome natural predators. In many cases, chemicals kill predators or drive them away. If you set out to build a chemical-free lawn that is rich in organic matter, you don't have to buy beneficial insects or microorganisms—they will flock to your lawn all by themselves.

Third, "bring on the birds." Question: What do birds eat? Answer: Bugs. Question: Why do birds migrate? Answer: Because bugs do. (Bugs migrate because of temperature and available food sources, too, but I think we're getting a little bogged down here.) Conclusion: Amtrak birds to your backyard so that they will eat their natural foods, which are the beetles and grubs that can damage your grass. Don't treat your lawn with chemicals, because birds have been known to have an aversion to them. Give birds a place to nest, water to drink, and food to eat, and they will be happy to live in your backyard. Look in the chapter "Top Ten Ways to a Beautiful Easy Landscape" later in this book, for a complete guide to attracting birds to your backyard.

Fourth, plant disease- and insect-resistant grass seed. As I mentioned earlier in this chapter, many of the new varieties of perennial ryegrass and fescue have been treated with a fungi that repels chinch bug and sod webworm. Most of these endophytic grass seeds are also very disease resistant. If your lawn is infested with diseases and insects, you may want to renovate it completely and replant with these new improved types of grass seed. You can read how in the chapter "Lawn Renovation, Restoration, Remodeling, and Rehabilitation."

Finally, treat your pests with some of the new biological pesticides now coming on the market. But please be careful. Some of these products, such as milky spore (*Bacillus popilliae*), have been of poor quality and have had to be removed from shelves.

Though they must be properly handled, beneficial nematodes, tiny parasitic worms, can be very effective in controlling most types of grubs. Neem, an extract from a tropical tree, is also being shown to be effective against sod webworms, billbugs, and cutworms. You can try to control the Japanese beetle population with pheromone traps and lures, and you will collect thousands of these critters each summer. But if your neighbors don't set up a Maginot line of defense right along with you, the beetles will simply fly in from somebody else's lawn and do their damage.

Even though some of the new biological pesticides are still working the kinks out, it is better to use them than the chemicals, which we know have problems. The best tip yet is to make your lawn so healthy and strong that a few bugs won't matter and diseases won't have a leg to stand on.

TIP NUMBER TEN: *Relax and Enjoy Your Lawn*

This is the fun part. I want you to relax and enjoy your lawn. Don't worry. Be happy. Weeds are not some alien beings from outer space that have to be hunted down with heavy artillery. A lot of people have just enough time to cut their grass and call it a day. They don't care if there are a few weeds.

Since I have been writing and speaking about gardening and lawn care, I feel it is important to make my lawn look good. But I have a little six-year-old son who now wants to play baseball, football, and other roughhouse games on the lawn. My two-year-old daughter wants to do everything he is doing, plus push her toy baby buggy all over the lawn. My wife wants lawn furniture out there. Finally, Flash, the horse who lives next door, got loose from his corral last month and stampeded through my backyard leaving huge hoofy divots in the lawn. That was the last straw.

At first I found myself yelling at them to stay off this part of the lawn and that part, too. Then I realized that's crazy. What's more important, my family or my lawn?

I've reached a compromise with myself. I've begun to overseed the lawn with turf-type tall fescue, which is more resistant to heavy traffic and wear and tear. I've decided to reduce the actual grassy parts of my lawn and plant larger areas with ground cover, flowers, and bushes. In fact, I'm now transforming my backyard into an environmentally sound, integrated backyard landscape. That's what I suggest you do, too. Forget about your lawn as grass growing on a military reservation. Start thinking about ways to make life easier for you and your lawn. Start thinking about your lawn as another room, an addition to your house, a wonderful fresh-air living space for you, your family, and your friends. Instead of furnishing it with sofas and chairs, plant flowers, shrubs, bushes, and ground covers and even a few trees. For a full-blown plan of how to make your lawn into a lawn when it isn't a lawn, read chapter three, "Lawn Renovation, Restoration, Remodeling, and Rehabilitation."

The Homeowner's Guide to Lawn and Landscape Tools

A lot of people really love their lawn and garden tools. They love to go to the equipment dealer's showroom and look at the new models of power mowers, shredders, chippers, and other equipment. They love to clean their tools, sharpen the blades, and take good care of their equipment.

On the other hand, a lot of people couldn't care less about their tools for the lawn and garden. They use them hard and put them away dirty. They let the blades go dull and seldom change the oil and gas in their power equipment. Then the next time they go to use their equipment, it doesn't work, or it doesn't work well.

I fall into the category between these two extremes, and I think most other people do, too. The big questions facing most beginning lawn keepers and gardeners are: What types of tools do I really need to buy? How much should I spend? How do I take care of these tools? and How do I use them?

That's what this chapter is all about. I am going to gather all the experience I have with tools, add to that conversations I've had with tool manufacturers and many other gardeners, filter out as much extraneous information as possible, and present you with a list of tools that I think you will need to take care of your lawn and landscape with a minimum of fuss and worry and causing the least amount of negative environmental impact.

This will be a rather opinionated guide to lawn mowers, chippers and shredders, rakes, spreaders, pruners, loppers, saws, shovels, spades, hoses and watering equipment, and other implements of destruction. I am not going to describe every single possible thing you can buy. Instead, I am going to advise you on things I think you should buy. If I don't talk about it here, it means that I don't use one and I think you probably don't need to use one, either. Please feel free to acquire any other tool you feel you want that I have not mentioned, however. For a complete guide to compost equipment, I refer you to chapter four of this book, "Clean, Easy Compost."

Lawn Tools

LAWN MOWERS

Riding Mowers. For a typical urban or suburban lawn, you really don't need a riding mower. If you feel you have a lawn large enough to warrant a riding mower, however, please get one that has a mulching blade or a mulching attachment. Leaving grass clippings on the lawn is one of the best things you can do for your lawn and for the environment. Otherwise, choose one of the following walk-behind mowers.

Reel Mowers. Many of us cut our first grass on an old-fashioned push-type reel lawn mower. The old ones were heavy and difficult to use. But now there are new ones, made by American Great States, that are made of steel alloy, which means they are lightweight and easy to use. They cost less than $100 and you never have to change the oil, buy gas, or worry about them starting. They are probably better for smaller lawns, although you can cut grass just as fast with a reel mower as you can with a power mower. You can burn over 350 calories an hour pushing a reel mower, so if you are looking for some quality exercise, this might be a good choice for you. Finally, running a power mower for thirty minutes emits the same amount of pollution as driving your car 170 miles. If you are concerned about air quality, a reel mower is the most environmentally sound mower you can buy. The only drawback of a reel mower is that the highest setting on the blades allows you to cut the grass at only 2¼ inches, rather than the 3 inches necessary to a successful environmental lawn-care program for northern grasses. Southern grasses can be cut shorter, so the reel mower might work better in certain areas.

Electric Lawn Mowers. In the past, the biggest problem with electric lawn mowers has been the long cord that seems to always get in the way. Now two new cordless electric lawn mowers by Ryobi and Black & Decker have liberated you from the cord. These are lightweight and easy to use. The Ryobi mulches the grass with its Mulchinator mower, and Black & Decker is introducing a cordless electric mulcher as well. Just plug the mowers into common household current to recharge their batteries. They will run for thirty to ninety minutes on a single charge, and they seem powerful enough to cut the grass very well. Cordless electric lawn mowers have an extra advantage for senior citizens who feel that pulling the starter cord on a power mower puts a strain on them. Electric mowers have push-button start, and they are also quite lightweight.

Gas-Powered Mulching Mowers. Mulching mowers are like giant food processors on wheels. They cut and recut the grass clippings until they are pulver-

Get a good lawn mower. Electric mulching mowers, power mulchers with a bagging attachment, or lightweight reel mowers all do a good job. Remember to leave your grass clippings on the lawn where they become a valuable source of free fertilizer and organic matter. Grass clippings do not cause thatch.

ized. The clippings are then blown down into the lawn, where they disappear and become a free source of fertilizer and organic matter. Most of the better-known lawn mower companies now make mulching mowers, including John Deere, Toro, Bolens, Troy-Bilt, Snapper, Cub Cadet, Simplicity, Honda, Ariens, Poulan, Husqvarna, and others. As a consumer, here's what you should look for:

1. a 4- or more horsepower engine

2. an enhanced cutting chamber

3. a good guarantee

Although it's not necessary, if you have trees that shed a lot of leaves in the fall, I think you should get a mower that has the capability to attach a collection bag, so that you can mow and shred fallen leaves that you can then put in your compost bin. This makes your mower two tools in one, a mower for grass and a shredder for leaves.

One of the perceived problems that some people have voiced about mulching mowers is that you have to cut the grass more often. It is true that a mulching

Power Mower Safety

Power lawn mowers can be dangerous. A mower blade can revolve about four thousand times per minute and hurl rocks or other debris up to two hundred miles per hour. The United States Consumer Product Safety Commission recorded more than seventy-six thousand injuries related to mowers in 1991. Thirty-five thousand people are treated in emergency rooms each year with lawn-mower–related injuries. Here are some precautions you should take before you start up your power mower again:

- Read the manual for any important safety instructions.

- Clear the lawn of rocks, sticks, or other debris before you mow.

- Wear leather shoes or at least closed-toe shoes, long pants, and a shirt when you mow. Never wear sneakers or sandals.

- Do not fill the gas tank when the mower is running or when the mower is still hot. Fill the tank outside rather than inside the garage or shed.

- Turn off the mower before clearing any clogs or clippings underneath.

- Do not mow when kids or pets are nearby.

- Mow across a slope with a walk-behind mower and always mow in a forward direction. Mow up and down a slope with a riding mower, so it won't tip over.

- Never, ever carry riders on a riding mower.

mower will not handle an overgrown lawn very well, especially one that is wet, but cutting your 4-inch grass down to 2 inches is terrible for your lawn anyway. For best results you should follow the one-third rule, i.e., never cut more than one third of your grass's height at any one time. Furthermore, cutting your grass with a mulching mower saves you time because you are not bagging and dragging those clippings off to the landfill.

Power mowers do emit more pollutants than either reel or electric mowers. California and soon many other states with large urban populations will be implementing restrictions on exhaust emitted from gas-powered lawn mowers. But most of the big manufacturers are rushing to redesign their engines to comply with air-quality standards, so you can ask about that when you buy your mower.

Power Mower Maintenance

To keep your power mower in tiptop condition, here's how to maintain it:

- Clean or replace the spark plug every year. A smoother-running engine burns fuel more efficiently and creates less air pollution. It also makes starting a mower a lot easier.

- Sharpen the blades at least once a year. Twice a year is really even better. Dull blades damage your grass, making it more disease prone, not to mention how awful your lawn looks.

- At the end of the cutting season, drain the gas and oil from the engine and put the mower in as clean and dry a place as you can find. Please read your owner's manual for directions on draining oil and gasoline. Oil and gas left in an engine deteriorate and can cause engine wear and failure the next year.

- Keep the underside of your mower clean and free of clippings and debris that can corrode the paint and damage the metal.

CHIPPERS AND SHREDDERS FOR LEAVES AND LIMBS

As I mentioned before, I prefer to use a mulching mower with a convertible bagging attachment to harvest leaves in the fall to add to my compost bin. The leaves are automatically shredded and caught in a bag that is easy to empty into the waiting bin. But if you have a reel mower or a dedicated mulching mower, you still need some equipment to harvest those leaves and shred them. Unshredded leaves take forever to break down and decompose. I like to harvest my leaves in the fall, compost them over the winter, and use them as mulch or dig them into my garden soil in the spring. Chipper-shredders come in all shapes and sizes. I've seen 8-horsepower models that can pulverize brickbats and 2-by-4 lumber. If you have a lot of timber on your property, and if you want to shred your Christmas tree, hedge clippings, and fallen branches, I suggest you get a large powerful shredder.

But most people only want to shred leaves, and for that purpose I can recommend an electric leaf shredder. You rake up the leaves, add them to the top of the shredder, and they come out the bottom all shredded up. They aren't too expensive, and they do a good job with a minimum of fuss.

SPREADERS

A spreader is a very useful tool for spreading fertilizer, grass seeds, lime, or organic matter for top dressing in a clean and easy fashion. Of course, you can always broadcast the materials by hand, but then you stand a good chance of getting uneven coverage and poorer results. Many companies make good spreaders. Look for one that is made out of heavyweight plastic or sturdy metal. It doesn't matter if you get a rotary or drop spreader, just make sure the spreader can distribute lime, seeds, and fertilizer and that it is adjustable to conform to various manufacturers' specified settings to get proper distribution of material. Also, be sure that the hopper is large enough for your lawn, so you don't have to keep refilling.

COMPOST BINS

Everything you ever wanted to know about compost bins is in chapter four of this book, "Clean, Easy Compost." The most important consideration is to get a bin and use it. You can then harvest lawn and leaf material, kitchen scraps, and household paper products like paper plates, napkins, and towels that will convert to $50 to $100 worth of mulch and organic matter each year. Get a sturdy bin with a tight-fitting lid measuring approximately 3 feet by 3 feet by 3 feet, or 1 cubic yard. It doesn't matter if the bin is made of wood, fencing wire, or hard plastic, just so it is sturdy, has louvers or slats for ventilation, and looks okay in your yard. Then, start filling it up with leaves, grass clippings, crop residue, and kitchen scraps (see page 66). Turn the material once a month, and you should get finished compost in six months to a year. Empty the bin and start all over again.

LAWN RAKES AND GARDEN RAKES

I have always been a fan of bamboo lawn rakes. I find that the teeth of the newer plastic rakes tend to break too often. My second choice is a metal lawn rake because it is durable and hard working, but I just like the feel and look of a bamboo rake. Also, a bamboo rake generally has a wider set of teeth, meaning you can rake up more leaves with one swipe. You will use a lawn rake to gather leaves in the fall and also to give your lawn a surface aerating during the spring lawn cleanup each and every year.

A garden rake, on the other hand, must have sturdy steel teeth, so that you can level out the soil where you are going to plant new grass seed or a new flower bed

and when you tamp the seeds down into the soil after they are planted. I have always used a flat-head rake, which attaches to the handle in a T shape, rather than the bow-head rake, which attaches to the handle by two bowed steel rods. I just find I can level soil better with a flat-head rake.

As with all hand-held tools, buy the type that is the most sturdy, that feels like it can take the most work and abuse. It will probably be the most expensive.

Get the right tool for the job. Professional quality long-handled tools will make lawn and landscape work joy, not drudgery. Here are a few (from left): Landscaper's shovel, garden rake, forked spade, bamboo leaf rake, metal leaf rake.

Be sure to clean and oil your tools after each use so they will be in good shape for you when you are ready to use them again.

Hand Tool Care and Maintenance

It's only important to take care of your tools if you want them to work well for you next time you need them. It is very important with metal rakes or shovels to clean all the dirt off the metal and then wipe the metal with an oily rag to prevent rusting. It is also a good idea to wipe wooden handles with wood oil once in a while to keep the wood clean and lightly coated so it won't dry out and crack. Finally, please designate places where your tools are hung up and put away and then put them away there every time. There is nothing more annoying than having to rummage around in your garage for a tool when you want to use it.

OTHER NEAT LAWN TOOLS

The basic tools for lawn care are a mower, a spreader, a leaf rake, a garden rake, a compost bin, and maybe a chipper or shredder. Of course, you need a hose and a sprinkler, but I am going to talk about that later in this chapter. But there are a few other interesting tools that you might want to consider.

Dandelion Digger. This is a small, flat, V-shaped metal head attached to a long wooden handle. You simply poke the head down under the dandelion, plantain, or other offending weed and pop it out. I prefer the long-handled tool to the short-handled one, because I think it is easier to use. These are widely available in hardware stores and garden centers.

Scythe and Sickle. A scythe is the long-handled tool carried by the Grim Reaper and Father Time and is actually a very useful tool for cutting weeds. Farm workers used scythes for hundreds, even thousands of years before the invention of the mechanical harvester. I am not a big fan of electric or gas-powered string trimmers, because they are noisy and I am kind of old-fashioned. I've been using my wife's grandfather's scythe (he was superintendent of grounds at the Sleepy Hollow Cemetery and certainly knew what he was doing), and I've become fairly adept at swinging the scythe and cutting weeds on our property. It's important to get a good-quality carbon steel blade, which should be kept sharp with a special stone you can buy along with your scythe. The shaft of a scythe is called a snaith, and the two handles are called nibs. I prefer a wooden snaith, which gets burnished with hand sweat and oil over a period of years, but aluminum ones are available, too.

I use a sickle, best described as a short-handled, curved-blade grass hook, to cut weeds close up to our rock garden that I can't get to with the scythe. This, too, works very well as long as the blades are sharp. It doesn't cost very much, it never breaks down, and I prefer the sound of the sickle and the scythe cutting weeds to the sound of the power trimmer doing the same job.

Landscape Tools

Even though your lawn is an integral part of your landscape, you will need some different tools to manage the shrubs and other plantings around your yard. To take care of the trees, shrubs, and flowers in your environmentally friendly, beautiful easy landscape you will need a couple of pruning shears, a shovel or two, hoses, and rainwater harvesting equipment, and a few extras that will make your gardening experience a lot more pleasurable.

Remember Mr. Natural, the Buddha-like character from the underground R. Crumb comics of the 1960s and 70s? One of his favorite sayings was "Get the Right Tool for the Job!" Nothing is worse than fumbling around with the wrong tool, especially in landscaping, in which the right tool can make your landscape beautiful and your effort a lot more rewarding.

My philosophy about landscaping tools, and most other tools for that matter, is to buy expensive premium-quality tools. Especially when you are just starting out. You really can't afford to buy cheaply made tools that not only will break and bend but also will

You need only a few well-chosen high quality hand tools to manage your lawn and landscape. Here are a few you will need (clockwise from bottom right): long-handled lopping shears, bypass pruning shears, scissors, leather holster for shears, forked trowel, digging trowel, fold-out pruning saw, sharp knife, and watering can.

do a poor job of cutting, digging, or whatever else you want to do with them. Your local better-quality nursery or garden center will be a good place to buy most of these tools. The manufacturer's brand names to look for include True Temper, Sandvik, Corona, True Friends, Felco, and Snap Cut. If you are having a hard time locating what you want, I suggest you try the A.M. Leonard catalogue, 6665 Spiker Road, P.O. Box 816, Piqua, OH 45356, or the Walter Nicke Co. catalogue, 36 McLeod Lane, P.O. Box 433, Topsfield, MA 01983.

PRUNERS, LOPPERS, AND SAWS

Hand Pruners. You are going to use a pair of hand-held pruning shears, a.k.a. secateurs, more than any other landscape tool to prune bushes, shrubs, hedges, and small branches up to ¾ inch thick on trees. You will typically use hand-held pruning shears to prune roses, azaleas, and hydrangea-sized bushes and thin sucker branches on fruit and ornamental trees. Look for professional-quality bypass pruning shears with curved steel blades and cushioned grip handles. Bypass pruners make cleaner cuts with a scissor action. Cleaner cuts not only leave you with a

better-looking shrub, they also help prevent diseases in your shrubs. You should avoid the anvil-type pruners, because they tend to crush rather than slice through the stems. You absolutely, positively must buy a leather holster for your pruners to keep track of them and for ease of use in the field.

Pruning helps make your trees and shrubs healthier by cutting out dead limbs or stems. Buy a top quality set of bypass-type pruning shears. Keep them sharp, cleaned, and oiled, and carry them in a holster that you can attach to your belt.

Lopping Shears. These are also known as long-handled pruners because they have the same bypass steel blades attached to wooden or aluminum blades 12 to 36 inches long. These are designed to prune or lop off branches up to 1½ inches thick.

Use lopping shears with bypass blades for larger jobs. Remember to prune spring-blooming shrubs just after their blooms have faded and summer- and fall-blooming shrubs in late winter before the new growth of spring.

You would typically use loppers on bushes the size of lilacs, rhododendrons, and flowering quince and for lateral branches on fruit and ornamental trees.

Look for lopping shears with curved forged-steel bypass blades and wooden handles that are not too heavy or too long for you to manage. Ratchet loppers are more suitable for professionals and are too heavy for backyard landscapers.

The same pruning tips and care go for lopping shears as for hand shears.

Pruning Saw. There are lots of different sizes of pruning saws, but I think the best one for you to begin with is the folding pruning saw. These saws have 7- to 10-inch blades that fold into a wooden handle much like a folding pocket knife. These are safe and easy to use, cutting branches up to 3 inches in diameter. The blades are very jagged and sharp, and the saw cuts on the pull stroke, which means it is less likely to bind up in the wood when cutting. Most of the quality companies listed on page 34 make excellent folding saws, but we should add the Fanno brand name for saws.

As you get more experienced and your trees get larger, you may find the need to get a larger, fixed-blade pruning saw, a long-reach pruning saw, or even a compound-action pole saw with sectional wood or fiberglass poles to prune limbs from trees you can't reach from the ground.

Pruning Tips

- Pruning helps improve the health and vigor of a plant. If a plant is all lanky and stretched out, it has to work too hard to support those outer extremities, causing it to weaken. A well-pruned plant will be more compact, with fuller growth. Pruning also increases flowering and fruiting of a plant.

- The best time to prune shrubs that bloom in the spring is immediately after they finish their blooming cycle. The plants produce the next year's blooms on new growth over the summer. If you wait too long, you inhibit the next year's flowers.

- Naturally, you would never prune a flowering fruit tree like apple or pear until after you have harvested the fruit. The best time to prune fruit trees, grapevines, and summer flowering plants like hydrangea and roses is in late winter or early spring before the sap rises and the plants begin to bud.

- You can prune obviously dead branches at any time.

- Make your cut at an upward 45-degree angle when you are pruning to the next bud or branch. When you are taking off a whole branch, make the cut as close to the remaining limb as possible to leave as little stub as possible.

- Keep your shears oiled at the spring and pivot point and have your blades sharpened at least once a year or whenever they get dull.

SHOVELS, SPADES, AND TROWELS

When buying shovels or spades, it is important that you get tools that are well made. Look for those that are called either solid-socket or solid-strap handled, because they can take wear and tear better than the cheaper tang-and-ferrule method of attaching the blade to the shank. A solid-strap handle wraps the wooden handle all the way to the grip with a solid piece of metal, or strap, which is attached to the blade itself for extra strength. A solid-socket handle, sometimes called half strap or three-quarter strap, also wraps the handle in metal but only about halfway up the handle toward the grip. You and your back are going to put a lot of pressure on these spades as you dig with them, and you don't want them to break. You can see how the steel strap will make for a stronger, harder-working tool.

Nursery Spade. The one tool you will definitely need in this category is the short-handled nursery spade. You will use this primarily to dig holes to plant small trees and shrubs and to separate and transplant perennials. But this shovel is a perfectly fine all-purpose tool to dig around in your compost bin and to dig out and shovel dirt, compost, and other organic matter around your landscape.

Look for a professional-quality steel blade about 7 inches wide and 10 inches long, flat or slightly curved at the bottom and with turned steps at the top. Some blades will be slightly tapered at the bottom for digging in tighter beds, and that is fine, too. Look for a solid wooden handle with a "D" grip. These spades are sometimes called garden spades, border spades, shrub spades, and even ladies' spades when they are lighter weight.

Forked Digging Spade. A forked digging spade looks like a giant table fork, with four steel tines attached to a wooden handle with a "D" grip. This spade does a better job of digging and breaking up clumps of soil in your flower, herb, or vegetable beds, but it's not very good for transplanting or digging holes. It can also be used to turn a compost pile and dig hay and manure barn sweepings—which most farmers will give away free—that you can use as mulch or to make compost. Look for the same full or part strap handle and high-quality metal blades as you would in buying your garden spade.

After using it, be sure to clean off any dirt or debris from the metal blades of your spade or fork. Wipe the blades with an oily rag to prevent rust. Occasionally wipe the wooden handle with some wood oil to keep it from cracking and to keep it clean.

Trowel. You are going to need a hand-held garden trowel to dig holes to plant and transplant bulbs, herbs, perennials, and potted annuals. Look for one that is made of heavy-gauge steel that is firmly attached to the wooden or comfort grip handle. Poorly made hand trowels will bend and break, leaving you without a tool just when you need it.

HOSES, RAIN BARRELS, AND WATERING CANS

All landscapes need water to survive. In ideal conditions rainwater would naturally come and satisfy all of your irrigation needs. That does happen in Ireland and Great Britain and in some parts of the Pacific Northwest. But most parts of North

America have periodic droughts, during which you have to supply the water. But water is a precious commodity. As our population grows, more demands are put on water, and we as gardeners and yard keepers need to do our best to make every little drop count. Luckily, there are some good tools to help us be water wise.

Soaker Hose. A soaker hose is one that weeps water out through its pores and delivers water slowly and as close to the root zone as possible. That's the most water-conserving and beneficial type of watering you can give your shrubs, flowers, and landscape. Soaker hoses come in 25- and 50-foot lengths, even longer in bulk if you want to install your own underground irrigation system. They feel a little more like foam rubber than regular hoses. You can simply install the soaker hose in your beds or around shrubbery, cover it with mulch, and leave it there. When you want to water, run your regular hose to the end of your soaker hose and turn on the water. A full explanation of soaker-hose techniques is included in the chapter "Top Ten Ways to a Beautiful Easy Landscape" later in this book.

Regular Hose. You will also need a regular rubber hose to water your lawn, hook up the soaker hose, and probably wash the car. Don't buy a cheap plastic hose. Buy an expensive rubber or reinforced rubber hose with solid-brass fittings that will last a lifetime. Pay more now but save more in the long run. Look for the most reputable brand names in quality hoses, including Moisture Master, Swan, and Gilmour.

Nozzles. I don't advocate watering your lawn or landscape with a hose and

nozzle. A soaker hose is by far the best way to do it. But from time to time you will have a need for a hose and nozzle, so I suggest you buy a solid-brass twist-type nozzle that is about 4 to 5 inches long. These things never break, they last forever, and they give you a spray from fine to direct and even shut off. A very good long-lasting buy.

Buy a good quality rubber or reinforced plastic garden hose. It will give you many years of faithful service. An all-purpose solid-brass nozzle costs less than $10 and will give a stream of water from a light spray to a hard flow with a twist of your wrist.

Sprinklers. You do need a sprinkler to water your lawn. Please read the watering section in the chapter "Top Ten Ways to a Beautiful Easy Lawn" for complete details. The bottom line is that you want to buy a solid-brass impulse or pulsating sprinkler, the kind you see being used on golf courses, rather than the overhead-spraying, oscillating sprinkler, the kind you see showering water on many suburban lawns. An impulse sprinkler just does a better job. There are lots of plastic impulse sprinklers that are quite easy to break, but I think you should buy a more expensive brass one that will last a lifetime.

Rain Barrel. Every time it rains, the roof of your house sheds several gallons of water down the gutters and downspout. You can harvest hundreds of gallons of that water over the course of the growing season by installing a simple rain barrel at the end of the downspout. A rain barrel can be as simple as a heavy plastic garbage bucket or a metal or wooden washtub. But you might want to consider a big 50- or 60-gallon hard-plastic barrel from the Great American Rain Barrel Co. in East Boston, Massachusetts. These are recycled olive and pickle barrels that are very

attractive, have specially designed lids to collect water and repel debris, have water-release valves and spouts, and can be connected to other rain barrels to really collect a lot of water. Once that barrel is filled with water, there is enough pressure to actually attach a hose to the spigot and water with that. Otherwise, you can dip the water out with a galvanized-steel, 3- to 5-gallon watering can with a metal rose sprinkling spout. This is a very effective and low-impact way to spot-water flowers in your landscape.

Place a rain barrel at the bottom of your down-spout and harvest hundreds of gallons of valuable precipitation over the course of the season. You can link together several barrels to capture a continuous supply. A large barrel like this one provides enough water pressure to allow a hose to be used with it.

Some of the other landscape tools that I find useful and fun to have include a good-quality pair of leather gloves to wear when pruning, handling shovels, and doing other work that will protect my hands from getting too many calluses and, more important, prevent deep cuts from thorns and prickly bushes.

Another tool you might consider is a wheelbarrow. It is very useful for hauling finished compost to the appropriate bed, for loading up your garden tools and planting supplies if your beds are a little bit of a walk from the house, for transporting buckets or watering cans full of water from your rain barrel to faraway beds, and other hauling chores. I like to use my wife's grandfather's old wooden wheelbarrow with slatted sides that can be removed to haul poles or other materials that don't fit in the usual compartment. I can also take those sides off and put the wheelbarrow in the trunk of my car if I need to use it at another location. Another good option is a two-wheeled garden cart that is easy to use and can haul a lot of material.

Finally, no gardener, lawn keeper, or landscaper would dare to venture out to the backyard without his trusty pocket knife. You can use it to harvest fruits, vegetables, herbs, and flowers, to prune slim branches on bushes, cut open bags of mulch, slice twine, and do dozens of other garden chores. I suggest you get a good-quality folding pruning knife with a single 3- to 5-inch carbon or stainless-steel blade with a wooden or plastic handle. Many garden knives have bright orange handles so that you can find them in the dirt after you have set them down.

Now, naturally, there are many more tools that you will buy or want to buy. This list gets you off to a good start. Remember, take care of your tools, and they will take care of you.

Lawn Renovation, Restoration, Remodeling, and Rehabilitation

T he 1980s were a very unusual decade for homes and lawn care. The baby boomers came of age and began to nest; they bought and built homes, they started families, and they took on the stewardship of their lawns. At the same time, the parents of the baby boomers started to retire. They had their homes and their lawns and their way of doing things, but once their grandchildren began to come over and play, they began to worry about the potential harm of chemicals on their lawns. Both of these huge groups of people began to ask how they could have good-looking lawns that don't require a lot of time and don't demand the use of chemicals.

Luckily, people's interest in the environment was reawakened in the 1990s, and a whole new wave of grass seeds, products, and techniques that make non-chemical lawn care a lot easier became more available. I hope I can bring some of this rapidly advancing school of environmental lawn care to you in this chapter.

Please read the first chapter in this book, "Top Ten Ways to a Beautiful Easy Lawn," before you start reading this chapter and reworking your lawn. It is important for your success to understand the philosophy and practice of environmental lawn care before you do the actual work of transforming your lawn from a weedy mess into a good-looking, but definitely low-maintenance patch of green. You might also benefit from reading the parts of the chapter "The Homeowner's Guide to Lawn and Landscape Tools" that pertain to lawn mowers before you get started, so that you can be armed with the very best tools for your uses.

Lawn care is actually very simple, and many of the things I will tell you in this chapter are not new. What is new is the attitude you need—I call it Nineties Lawn Consciousness—which combines modern science, low-maintenance techniques, improved grass seed varieties, and a willingness to trust in nature, so that you can have a beautiful lawn that is easy to maintain without the use of chemicals.

After a little bit of history and interesting tidbits about lawns, this chapter will tell you how to make a lawn assessment, how to shop for the new improved grass seed varieties, how to patch small problem areas, how to change your lawn without digging the whole thing up, and, of course, how to plant an entirely new lawn.

A Short History of Lawns As We Know Them

Even though there is some speculation that patches of grassy areas may have been cultivated in ancient Roman gardens, it wasn't until the rise of the bourgeois upper-middle-class country gentry in the seventeenth and eighteenth centuries that the concept of a lawn became imaginable. Even then, the lawns around the houses were probably meadow mowed by sheep as part of their grazing land. Lawns then were really meadows beginning to become lawns.

As the sheep were moved away from the home and ornamental gardening and landscaping became more common and popular, the lords and owners began to maintain their lawns by having servants cut the grass with scythes. Of course, labor was cheap in those days. (Come to think of it, I once had a roommate in Columbia, Missouri, who cut my lawn with an ordinary $5 weed whacker in the days before I owned a lawn mower.)

The first real watershed of modern lawn care came in the early 1830s when Englishman Edwin Budding invented the first mechanical lawn-cutting machine, which was manufactured by the Ransomes Company. The first ones were push-powered machines, followed by some designed to be pulled by horses, followed by all types of variations until 1902, when Ransomes unveiled its first gasoline-powered lawn mower. People continued to use nonpowered push mowers into the 1950s, when the rotary power mowers we know today became common and affordable. Now, with our concerns about air pollution, power mowers are being modified, reel mowers are making a comeback, and electric mowers are beginning to get noticed.

A Look at Today's Lawn

A lawn is more than a perfectly manicured organic carpet that we use to cover the ground around our homes. A lawn is a very special type of garden. A typical 1,000-square-foot lawn is a collection of over a million tiny grass plants. Each grass plant competes with other grass plants as well as weeds for food and water. Each plant needs good soil to grow in. A good-quality soil is just as important for healthy grass as it is to grow crops on farms. In fact, lawns are really a type of perennial grassland agriculture.

Your lawn is actually a well-balanced, integrated ecosystem. As explained in chapter one, each grass plant is made up of the tall green leaves that we see above the ground and a deep, growing root system that digs down below the ground. The root system should be as deep below the ground as the grass leaves are high above the ground. The green leaves above the ground need nitrogen. The grass clippings themselves, if left on the lawn, will decompose and provide more than half of the nitrogen fertilizer your lawn needs.

The deep roots below the ground need fertilizer high in phosphorus. Tree leaves, if minced with a lawn mower, can provide your lawn with a good dose of

---

According to Cornell University and the National Gardening Association's annual Gallup survey:

- There are twenty-five million acres of lawn turf in the United States. Five million of those acres are golf courses, cemeteries, athletic fields, and other open spaces. The rest is managed by individual homeowners.

- Sixty-one million American households participated in lawn care in 1990. Retail sales of lawn-care products totaled $6.4 billion in 1990.

- In 1990, we spent $51.2 million on hoses, nozzles, and sprinklers to water our lawns and gardens.

- We spent $19.6 million on grass seed alone in the same year.

- Twenty-nine percent of all lawns are treated by commercial lawn services. Homeowners care for 70 percent.

- The average homeowner spends $200 per year to maintain the lawn, covering costs for a lawn mower, gas and oil, fertilizers, and other items.

---

phosphorus. If you simply mow your grass and leave the clippings on the lawn, you will be providing most of your lawn's nitrogen fertilizer needs. If you mow a few fallen leaves into your lawn, you will be providing it with most of its phosphorus fertilizer needs. If you scatter some of your fireplace or wood stove ashes over the lawn, you will be providing most of its potassium fertilizer needs. It's all really very simple and integrated. Nature has provided us with a synergetic ecosystem if we just don't mess it up.

Your lawn is alive. Think of it as a wild jungle teeming with animals. In this case, the animals are worms, insects, and millions of microorganisms. A healthy lawn is full of animals, all digesting nutrients and fertilizing the grass. A healthy lawn is possible only if it is growing in a healthy soil. Just like good garden soil, good lawn soil needs to be rich in organic matter, crumbly, friable, well aerated, and well drained.

A lawn is also an important air-pollution purifier. The grass leaves inhale carbon dioxide and other greenhouse gasses and exhale oxygen, which we need to breathe to stay alive. A lawn measuring just 50 feet by 50 feet releases enough oxygen to meet the needs of a family of four. Like trees, lawns are a natural form of air conditioning, because they absorb heat and give off cooling water vapors.

A lawn is also a first line of defense against soil erosion. A healthy lawn prevents heavy rain waters from washing away topsoil and depleting the natural resources we need to grow food to feed ourselves.

Many garden writers like to make disparaging comments about lawns, how ugly they are and how wasteful they are in terms of water and fertilizer use. But if you follow the environmental lawn-care program outlined in this book, your lawn will not be wasteful, and you will be providing a lovely accent to your ornamental plantings as well as a wonderful secure place for your children and grandchildren to play.

Getting Started

SURVEY YOUR LAWN AND TAKE NOTES

A healthy soil grows a lawn that is beautiful to behold and easy to manage without the use of chemicals. The first step on the path of building a healthy lawn is a lawn tour survey. Grab a note pad and a pencil and walk around on your lawn. Once you have gathered all of the following information, you can follow the tips I have outlined for you in this chapter, or you can have a long talk with your county Cooperative Extension agent or lawn-and-garden center professional to decide which grass types you should be growing, how much seed or fertilizer you need to buy, and if you need to make major changes like removing trees or rerouting your driveway. Here is a list of things to take into consideration:

- How big is your lawn? You need to know the square footage, so you can determine how much grass seed or fertilizer you might need to buy. Either measure it with a measuring tape or mark it with strides, estimating that each stride is nearly 1 yard long.

- What type of grass or grasses are growing there? Dig up a small patch and take it to your Cooperative Extension agent, Soil and Water Conservation Service, or lawn-and-garden store. Most people in your county, state, or region are probably growing the same type of grass, so this is a simple determination for an expert.

- Look at the lawn. Is it thick and lush, or is it brown and spotty? Do you have a lot of weeds? What type? Dandelions? Are there brown spots or bare spots? Make notes of all these problems.

- What does it feel like to walk on your lawn? If it is soft and spongy, if it feels like you are walking on foam rubber, it probably means that you have thatch buildup that must be corrected. If it is very hard and bumpy, it may be heavily compacted.

- What sort of drainage does your lawn have? Do you have standing puddles of water after a heavy rain? Where? If there is a stream nearby, does it often flood your lawn? What about runoff from the street or gutters. Does it leave puddles or gashes in your lawn?

- Does your lawn harbor a septic tank drain field? More than 25 percent of American homes have septic systems. The drain field can cause depressions in the lawn. If it is not working well, it can contribute unwanted moisture and fertilizer to the lawn, also.

- Is your lawn in sun or shade? Are there trees growing overhead? Do you have an open sunny lawn, or is the light filtered by trees or shrubs? What types of trees and shrubs are growing there? How well does the grass grow near them or in their path?

- How are you going to use your lawn? Will the kids be playing games out there? Do you want to set up a table and entertain?

- Does your lawn have good air circulation? Poor air circulation can cause disease problems. Is your lawn fenced in by trees, wooden or stone fencing, or bushes? Does it get cooling breezes once in a while, or does it always seem kind of stuffy out there?

- Where do the driveway, garage, and walkways meet your lawn? How is the grass doing in those spots?

Make a general assessment of your lawn. If it is at least 50 percent green and healthy looking, with only a minimal infestation of weeds and patchy spots, you will probably want to follow my instructions for rehabilitation and restoration: overseeding (page 57). If you feel your lawn is too far gone, you will probably want to proceed to a full remodeling and renovation: starting a whole new lawn (page 54). Remodeling and renovation take quite a bit more work and you must keep the kids off the lawn for several weeks, but the results are wonderful and you can have a fabulous lawn starting from the ground up. In either case you will need to determine the right grass seed for your lawn and make sure your soil is properly balanced.

CHOOSE THE BEST KIND OF GRASS SEED

Chances are you didn't plant the grass seed growing in your lawn. Somebody else planted it thirty or forty years ago when it was built, or a contractor planted it when he built your new home, or it was simply the meadow grass that was there if you live in a Victorian, Colonial, or some other yesteryear home. That means you could either have grass grown from cheap, crummy seeds, grass grown from seeds that are way out of date, or simply the wrong kind of grass growing in your lawn.

If you want a beautiful lawn that is easy to maintain without the use of chemicals, you have to grow the modern, scientifically improved types of grass seed that are now on the market. Grass seeds today are much more disease resistant, some are even pest resistant, they have improved vim, vigor, and vitality, and they can even withstand drought and stress better than grass seeds we commonly planted just twenty years ago.

One of the biggest mistakes that people make in growing a lawn is trying to grow the wrong kind of grass in the wrong geographic place. Like most plants, certain grasses grow better in certain areas than others. Growing the wrong grass can cause a lot of environmental damage and doom your lawn to failure.

For instance, Kentucky bluegrass is a beautiful and well-known grass type that grows well in cooler parts of the country. But too often people who move from, say, Columbus, Ohio, to Scottsdale, Arizona, think that they can grow Kentucky bluegrass in the arid Southwest. That grass can only be grown in that location with an artificial life support system. You'll be using too much water, too much fertilizer, and too many pesticides. It is important for the success of your lawn to grow the right kind of grass for your location.

There are dozens of different types of grasses you can put on your yard to make a lawn. But all of these grasses boil down to two main types: cool-season grasses, which grow well across the northern and midwestern states, and warm-season grasses, which grow better in the hot and humid southern states. Cool-season grasses grow better in spring, early and late summer, and fall, when the weather is cool. They go dormant in the cold winters and during the heat of the summer unless they are irrigated. Warm-season grasses grow best during the heat of the spring, summer, and early fall. They go dormant during the slightly cooler late fall, winter, and early spring months of the southern United States.

Cool-Season Grasses for Northern and Midwestern Lawns

Kentucky Bluegrass. This is the most popular of all grass seeds and is very well suited for the Northeast, the Midwest, and the Northwest. It is a fine-blade, deep green lawn grass. It germinates within two weeks and will fill in your entire yard within two months. It recovers quickly from stress such as drought, pest damage, and weather extremes.

But, Kentucky bluegrass demands a lot of fertilizer, it doesn't like shade, and it is susceptible to drought, disease, and insect problems. The days of an all-Kentucky-bluegrass lawn are over. Now it is better if you use Kentucky blue as only 10 to 30 percent of a mixture, along with ryegrass and fescue. Look for named varieties such as Adelphi, America, Eclipse, Princeton, and Bonnieblue.

Fine Fescue. Also known as creeping fescue, red fescue, chewing fescue, and hard fescue, this is a fine-textured grass with leaves that resemble soft pine needles. It does well in less fertile soils and in shade. It is a good low-maintenance grass to grow because it prefers neglect and does poorly if given too much fertilizer and pesticide. Look for named varieties such as Reliant, Scaldis, Longfellow, Agram, Shadow, and Waldina. SR 5000 and SR 3000 are two fescues that contain pest- and disease-resistant endophytes. (See page 18 in chapter one for a further discussion of endophytes.)

Improved Tall Fescue. Older types of tall fescue, like Kentucky 31 and Alta, were once considered weeds in a fine lawn. But the newer improved varieties are easy to grow, and they withstand lack of fertilizer, compaction, and drought. They don't mind a little bit of shade, and they are so rugged that they are being used more and more on athletic fields, because they can take wear and tear. Tall fescue is also increasingly available with endophytes, and there is some evidence that it is not damaged by grubs in a lawn. Tall fescue can make up 60 to 80 percent of an all-purpose lawn seed mix. Look for names like Titan, Shenandoah, Cochise, Guardian, and Hubbard 87.

Perennial Ryegrass. Also known as turf-type ryegrass to differentiate it from the common annual ryegrass that is often grown in pastures or as a winter cover crop, perennial ryegrass is similar in appearance to Kentucky bluegrass and is very resistant to pest and fungus damage. It is really the grass seed of the future, because so much of it is now treated with endophytes and so much of it is being bred to be resistant to diseases such as brown patch. Look for named varieties such as Enviro, Turf Alive, Repell, Citation II, Commander, Pennant, Regal, and Sunrise.

White Clover. Growing white clover is actually one of the best things you can do for your lawn. Adding a small amount of white clover, no more than 5 percent, can make your lawn more fertile, because clover attracts nitrogen from the air and brings it into the soil. It is also very tolerant of drought, poor soil, and abuse.

Seed Mixtures. Good grass seed is more expensive than poor grass seed. In general, more expensive grass seed contains more of the named grass seed and less of the chaff, weeds, and filler that can accompany a grass seed package. Read the label for exact content of how much of the grass you want is in the package you are buying.

It is a good idea to plant a grass seed mixture of, say, perennial rye, fine fescue, and a little bit of Kentucky blue, because a combination of grass seeds is better able to withstand the ravages of drought, disease, or pests. If one type of grass in your lawn is being damaged, the other types may fill in and keep your lawn alive.

Most grass seed today is a mixture of several types of seeds. The packaging will almost always tell you what types of grass seeds are inside the box, whether the mixture is good for sun or shade, whether it will withstand traffic, whether the seeds are disease, drought, and pest resistant. When in doubt, ask for help. You tend to get a more informed sales staff at established lawn-and-garden centers and nurseries, but many of the chain home stores are training their staffs to know more about grass seed.

Warm-Season Grasses for Southern and Southwestern Lawns

Warm-season grasses seem to grow better below a general dividing line that runs from Atlanta to Birmingham to Dallas. North of that line cool-season grasses, especially those perennial ryegrasses and fine fescues treated with endophytes, should be tried and encouraged. The problem is that Bermuda and zoysia grass do not like to share a lawn with other types of grass. If you are having good luck with your Bermuda or zoysia grass lawn, stick with it. If you are having trouble, think about the new endophytic grass seeds as an alternative.

Bermuda Grass. This is the most popular and most widely grown grass in the South. It is more difficult to establish from seed than bluegrass, and often people will resort to sod to get a lawn of Bermuda grass going, although seeding works in most cases. Some Bermuda is best planted from sprigs or plugs. Once it is established, it fills in very quickly. It is very drought tolerant, and it doesn't mind salt sea spray for those living near coastal areas. One of the problems with Bermuda is that it does not like shade, which is a wonderful thing to have around the house during those hot southern days. Look for improved Bermuda names such as Cheyenne, Turcote, Turftex 10, Tifway, and Tifway II.

Zoysia Grass. This grass is very difficult to get established and very slow to grow and fill in an entire lawn. Many people plant sprigs of grass called plugs with the addition of tall fescue to get the lawn off to a good start. The main problem with zoysia is that it turns brown early in the fall and stays brown until late in spring. But it has many fine attributes. It is drought and disease tolerant, it doesn't demand a lot of water or fertilizer, and it doesn't need to be mowed as often as other grasses.

Bahia Grass. This is the type of grass to grow if you live in Florida, because it tolerates the sandy soil there but doesn't grow well anywhere else. It is very low maintenance and requires little mowing and little fertilizer. You really need to keep your mower blades sharp, however, because Bahia is very coarse textured. Argentine is the seed variety that gives the nicest lawns.

Centipede Grass. This is an outstanding low-maintenance choice for Deep South lawns. It will grow in acidic soils, it will tolerate shade, it requires little fertilizer, and it grows so slowly that you won't have to cut it very often.

St. Augustine Grass. Also known as Charleston grass, this is a rather coarse-textured grass that will live in light shade and produce a deep green along the Gulf and south Atlantic coasts and as far north as Atlanta and Birmingham.

Special Grasses for Western and Southwestern Lawns

Buffalo Grass. This is a low-maintenance gray-green grass that withstands both heat and cold, requires little mowing, fertilizer, and water, and is very resistant to insects and diseases. It is, however, rather clumpy growing, and it doesn't do well in shade. But new improved seeds and sprigs are now being developed; look for Prairie, 609, and Bison. Buffalo grass is an outstanding option for people who live in Texas and other parts of the Southwest.

Blue Gramagrass. This is another prairie grass that forms a nice-looking dense lawn. It is not widely known yet; research is still being done to make it better for lawns, but it shows promise.

TAKE A SOIL TEST

Imagine this: Your child walks in the door and says, "Mom/Dad, I feel sick." You say, "What's wrong?" He/she replies, "I don't know." Do you then respond, "I'm going to shave your head and rub you all over with crushed moth ball?" Of course not! You take the tot's temperature and discuss the symptoms with the pediatrician, who may want a blood test if the disease seems serious.

A soil test for your lawn is the same first line of diagnosis for problems as taking a temperature is for a child. It should be done whether you plan to redo your lawn from scratch or simply rehabilitate it by overseeding.

At least half of your lawn's problems, including weeds, lack of fertilization, drought-stricken grass, and more, are caused by soil that has a pH balance that is either too acid or too alkaline.

The symbol pH stands for "potential hydrogen," and it is a measure of a soil's acid and alkaline levels. A lawn grows best when the pH level is between 6.0 and 6.8. A soil that is high in acid will have a pH below 6.0, and you will need to add lime to bring it into balance. A soil that is high in alkaline will need to have horticultural sulfur, gypsum, or peat moss added to bring it into balance. Rhododendrons and azaleas like an acid soil, but lawns like a more neutral soil, balanced between acid and alkaline.

Any Cooperative Extension agent and many lawn-and-garden centers can help you get a soil test done. Look not only for pH but also for the type of soil you have—gravelly, sandy, or hard clay—and whether or not it needs to have organic matter added to it.

A TIME TO PLANT

Once you know what type of grass seed is best for you to be planting, and you have taken a soil test to determine if you need to add any lime or sulfur to correct the pH balance, you are ready to plant. But when?

The very best time to plant cool-season grasses, either for overseeding or for an entire new lawn, is in the late summer or fall, six to eight weeks before your first expected frost. Spring is a risky alternative, because you are asking your expensive grass seed to compete with crabgrass seed, which is primed and ready to take over your lawn.

The very best time to plant warm-season grasses is in the early spring and summer, from March through June, when the soil has warmed to a temperature above 70 degrees.

Since there are so many different regional climates and even microclimates all over the United States, I suggest you contact your Cooperative Extension office or lawn-and-garden center professional for more specific grass planting dates in your area.

I realize you might not want to wait until the appropriate time to plant your grass seed, but you will be much better off if you do, because you will have a much better chance for success. In the meantime, follow some of the new improved fertilizing, watering, and mowing techniques that I mentioned earlier in this book to get your lawn in good shape. Who knows, you may find out you don't need to take such a radical step.

Rehabilitation and Restoration: Overseeding

If you have decided to overseed your lawn, it means that you believe you have a good base to start with. I call this method rehabilitation and restoration rather than simple overseeding, because you are going to build the soil for guaranteed future success rather than scatter a few seeds over the lawn. Remember, late summer to early fall, six to eight weeks before your first expected frost, is the best time to do this for cool-season grasses. Here's how:

1. Be sure you have had your soil test done and any lime or horticultural sulfur added to neutralize the pH balance several weeks before you start this process.

2. Cut the grass as low as your mower will go without hitting the blades on the dirt. Remove all the clippings. This is a great time to use a bagging mower. If you haven't used any chemicals on your lawn, you can place all the clippings in your compost bin.

3. Apply a grass-starter natural organic fertilizer—a fertilizer very high in phosphorus—or bone meal, which is naturally high in phosphorus, at rates recommended on the bag, usually 25 pounds per 1,000 square feet. Phosphorus helps grass roots grow strong, and that is what the grass plants are trying to do right now.

4. Rough up the soil with a power rake or a lawn dethatcher, which you can rent at a rental place. Or you can do this by hand with a sturdy steel garden rake or a steel dethatching rake if your lawn is not too large. Then use the rake head, either power or manual, to dig shallow, $\frac{1}{4}$-inch-deep channels in the soil so that the grass seed can nestle down in there like corn seeds in a furrow.

5. Sow the seeds at the rate recommended on the seed box label for overseeding existing lawns. This is usually 25 to 50 percent more seed than you would use for starting a whole new lawn. You need to plant more seeds when overseeding because the germination rate is not as good as it is when planting a whole new lawn. Spread half the seeds walking on one direction, using a seed spreader or broadcasting them by hand, and then sow the remaining half in a direction perpendicular to the first pass.

6. Gently rake the lawn again, this time to fold the seeds into the little channels you dug with the rake earlier. This helps get the seeds in good contact with the soil.

7. Top-dress the area with $\frac{1}{4}$ inch of topsoil, screened compost, dehydrated manure, or peat moss, lightly covering the seeds to prevent them from drying out or getting eaten by birds. Just fling the material over the lawn with a shovel and try to cover all areas with a light dusting.

8. If your lawn is large, you might want to rent a lawn roller and roll your lawn. If it is small, simply walk the lawn and tamp down the area with the flat head of your rake.

9. Water the area immediately and continue to water it lightly every day that it doesn't rain until the seeds germinate, anywhere from seven to fourteen days for perennial ryegrass to twenty to twenty-eight days for Kentucky bluegrass. Then reduce the watering and let the fall rains take over.

10. Let the grass grow slightly above its desired height before you give it its first mowing.

If you don't want to replant your entire lawn, you can still renovate it by overseeding—planting new, improved grass seeds right into your existing lawn. Cut the grass as short as possible; rake the lawn to scratch the soil and prepare a planting surface; sprinkle on the new grass seed; cover with a light coating of compost, peat, or topsoil to protect the seeds; and keep it moist until the seeds germinate.

SPOT PATCHING

You may find that even after you've planted a whole new lawn or overseeded an existing one there still might be a few small patches of your lawn where the grass just didn't get established very well. Or maybe you had to dig up part of your yard for septic repair or sewage work, or you removed some bushes and now you want to plant grass. There are many scenarios in which spot patching is necessary.

The one problem you often face with spot patching is that you can't always choose the time of year when it has to be done. If you can wait until late summer/early fall in northern areas or late spring/early summer in southern areas, please do. You always get better results if you plant at the appropriate time of year. But if you can't wait just follow these directions and you will still get good results.

Spot patching is a lot like overseeding except the area is smaller. You still need to buy a premium quality grass seed that is right for your climate and area. Dig or rake the surface of the ground to loosen the soil and remove any debris. Add a small amount of natural organic fertilizer, about 2 cups per 25-square-foot area, and a

$^1/_2$-inch-thick layer of natural organic matter such as compost, dehydrated manure, peat, or even top soil or potting soil if the area is small. Rake or dig the area again to incorporate the organic matter, the soil, and the fertilizer. Rake smooth.

Sprinkle on the grass seed thick enough to cover the area but thin enough so that the seeds are not on top of each other and that you can still see the earth below. Tamp the area with a rake and cover it lightly with straw or peat moss. Water the area lightly to keep it moist but not soggy. Keep the kids and pets off the area until the seeds germinate and the grass is growing well. Mow the patched area when it has reached the same height as the rest of your lawn grass.

Remodeling and Rehabilitation

After you have taken your lawn tour, you may decide that a complete lawn make-over is necessary. You may want to start from scratch because your home is newly built or because you have landscaped and added topsoil to the lawn. Or you may just be an energetic person who enjoys working around your yard and you want to rebuild your lawn to make it environmentally friendly from the ground up. Here's how:

1. As for overseeding, be sure you have had the soil's pH balance tested and added any lime or horticultural sulfur to bring it to a neutral balance between 6.0 to 6.8.

2. Clear and clean all the debris out of the lawn. Look carefully for any construction debris, especially pieces of wood that may rot and attract fungus and scraped-off paint chips that may be loaded with lead. Remove any large rocks and stones from the surface to prevent lawn mower damage and patchy grass growth.

3. At this point you must decide if you are going to kill off the remaining grass or weeds growing in your lawn or simply till the area and rake out the grass and weed debris. To kill the existing growth, most lawn services will suggest that you spray the lawn with glyphosate, the synthetic chemical herbicide found in Roundup. Glyphosate will kill your lawn with no doubt, but there are alternatives. Simply spread heavy-gauge black plastic over the lawn and leave it there for a week or two until the grass underneath is dead.

4. Dig up the lawn with a rotary tiller to a depth of 6 inches. You really want to break up any hard ground and sod to create a crumbly planting soil. You also want to be sure to grind up any remaining grass, sod, and weeds. These will become

(Above left) To start a whole new lawn, till up the area with a rotary tiller to a depth of 4 to 6 inches. Rake out any unwanted debris, grass plants, and weeds. (Above right) Add a 1- to 2-inch layer of compost, composted manure, peat, or other organic matter to enrich the soil and make it more crumbly. Add a natural organic fertilizer to feed the new grass shoots, and till the area again to incorporate this new material.

good fertilizers and help create a soil high in organic matter. A hand-held rototiller can be rented by the day or the week, or you can hire a professional by looking in the Yellow Pages or the classified section of your newspaper. You should rake out any remaining grass debris to give your new grass the best chance to get established and to reduce the amount of unwanted weed seeds.

5. Now is the time to add a natural organic fertilizer that is blended to contain an extra amount of phosphorus, which is beneficial to the establishment of the grass plants' root systems. You may also want to add bone meal, an important source of phosphorus, at the rate of 25 pounds per 1,000 square feet. Simply broadcast these elements or spread them with a drop spreader.

Smooth the area with a rake to create a level planting bed, free of rocks and debris.

6. Add a 1- to 3-inch layer of compost or other organic matter to the soil. Compost, commercially prepared dehydrated manure, and composted manure, are the very best materials to add. The regular barnyard manure that you might find at a local farm would ordinarily be good organic matter, but it usually has a high weed-seed content, and you want to avoid weed seeds at all costs when establishing a new lawn. You may want to contact your local lawn-and-garden center or nursery to see if they have truckloads of organic matter that they can deliver to you at a reasonable price.

7. Till the lawn again, incorporating the fertilizer and organic matter into the top 6 to 8 inches of soil.

8. Smooth out the soil with a rake, making the ground as level as possible. This may take a while, but it will prevent water damage in the form of puddles or rivulets caused by runoff.

9. Water the ground thoroughly to provide a damp but not muddy field for the grass to germinate in.

10. Sow the seed with a hand sower or push spreader. Different grasses have different rates of application, so read the instructions carefully. Be sure to choose the right grass for your area.

Fill the spreader with some of the new disease-, drought- and pest-resistant grass seed. Read the package instructions and set the spreader to apply the seeds at the proper rate. Be sure to get plenty of seeds on the planting area.

Tamp the entire area with a rake to help the seeds make good contact with the soil.

11. Rake or roll the sown area lightly to cover the seeds and make sure they have good contact with the soil. Tamp the area down or roll it lightly to form a good growing structure.

12. Since you are starting a whole new lawn, you will want to cover your grass seed with a light mulch. Straw—not hay, which has weed seeds in it—makes a good mulch because it protects the grass sprouts from drought and wind. It also decomposes and disappears into the lawn, adding more organic material to the soil.

13. Continue to lightly water the seeds every day until they germinate, which can be anywhere from seven days to three weeks.

Spread a thin layer of weed-free straw over the planted area to provide cover for the emerging seeds and help retain moisture.

LAWN RENOVATION, RESTORATION, REMODELING, AND REHABILITATION

Water the newly seeded area every day, keeping it moist until the seeds germinate and you see lovely green blades of grass peeking out from the straw. If you're lucky, rain will do this for you.

14. Begin cutting the grass once it has reached its desired height, from 2 inches for warm-season grasses to 3 inches for cool-season grasses.

SOD, PLUGS, AND SPRIGS

Sod is like a large carpet of grass that you place on your prepared lawn. It costs a lot more than grass seed, but you get a beautiful lawn overnight. The only drawback to sod is that not much of it is grown with the new endophytic perennial ryegrass or fine fescue. It features mostly Kentucky bluegrass. Ask for the new grass-seed varieties if you want to plant sod. A lot of southern grasses produce better lawns quicker with sod than with seeds.

When you plant sod, follow all the directions for soil preparation and testing, tilling in organic matter and fertilizer, smoothing the soil surface, and then simply lay the sod out as if you were laying carpet. Water the sod every day for the first two weeks and then slow down to every other day for a week or so until it gets established.

Plugs are small pieces of sod, and sprigs are pieces of turf that you can use to plant your lawn, which are a little more economical than planting whole layers of sod. Prepare the soil as you would for seeds or sod, and then plant the sprigs or plugs in a checkerboard pattern across the lawn, leaving every other square open. They take a little longer than sod, but they will fill in a lawn in less than a year.

~

Whether you overseed, plant a whole new lawn, plant sod, seeds, sprigs, or plugs, follow the lawn maintenance tips in chapter one, "Top Ten Ways to a Beautiful Easy Lawn," and you should have a beautiful green lawn that is weed and pest free for many years to come.

Clean, Easy Compost

Compost Happens

Compost happens. It just does. No matter how hard we try to stop it. In nature, the process of plant and animal matter decomposing and being transformed into elements of soil from which new plants grow goes on endlessly. The Bible is really offering a discourse on composting when it talks about ashes to ashes and dust to dust. Time passes. Compost happens.

But none of us lives in the Garden of Eden anymore. Most of us live in cities or in suburbia, where the idea of a pile of rotting, decaying matter in our backyards is abhorrent, not to mention illegal. The kind of composting I am talking about here is *controlled* composting for the backyard. We want to enclose yard waste and certain kitchen scraps in a tidy, functional, and attractive box that will produce finished compost in a reasonable amount of time—less than a year. Then we can empty the bin and refill it with leaves, palm fronds, grass clippings, dead flowers, mushy mangoes, tumbleweed, or whatever other debris ends up in our backyards every year.

In this chapter, I'm going to outline for you how to make goof-proof compost. We're going to debunk the myths and misinformation about compost, take you on an opinionated survey of compost bins, and help you choose the right bin for your needs and your pocketbook.

You may be asking yourself, "Why is there a chapter on composting in a book on lawn care and landscaping?" The answer is simple: Composting is the one activity that ties the lawn and the landscape together in a continuous cycle of growth, recycling, and reuse. Let me give you an example from my own backyard. Last year I made compost from the leaves that fall from my two huge maple trees along with spent flowers, some grass clippings, my shredded Christmas tree, and kitchen scraps, some of which were generated in my own herb and vegetable garden. This year I replanted my front lawn using my homemade compost as the organic matter that's added to increase the soil's fertility. I've also used my homemade compost to plant a row of mock orange bushes along my driveway and as a mulch on my perennial beds.

You see, compost is the glue that binds lawn and landscape together. I think everyone who has a yard featuring lawn grass, trees, shrubs, and flowers should have a compost bin. Here's how you can do it, too.

HISTORY OF COMPOSTING

Composting, or, more correctly, the recycling of crop residue, livestock waste, and food scraps, became widely prevalent as soon as civilizations became more settled, adopted agriculture, and stopped being wandering nomads, hunters, and gatherers. Once people stayed in one place and grew crops on the same ground year after year, they found it necessary to replenish the organic component of their soil to keep it fertile and productive. That's not to say they built tidy little compost bins. They simply spread the animal and plant waste over their fields and tilled it in.

Modern backyard composting was formulated by Sir Albert Howard, the British horticulturist who, in the 1940s, designed a wooden compost bin, which was then built by an Auckland, New Zealand, gardening club. Sir Albert's box measured 4 feet by 4 feet by 4 feet, which is a good size for a compost bin today.

The science of making compost and the importance of using it as a foundation of modern organic gardening was furthered through the work of Rudolph Steiner, the German philosopher whose concept of biodynamic gardening is still popular in Germany and at several Waldorf Farms in North America. J.I. Rodale, founder of *Organic Gardening* magazine, and Sam Ogden, author of the book *Step by Step to Organic Vegetable Growing* (Rodale, 1971), were also instrumental in the modern evolution and popularity of composting.

Today there are an estimated three million backyard compost bins at work in the United States and Canada. Even Martha Stewart, the glamorous author and television personality, makes compost at her Connecticut estate. According to *USA Today,* President Bill Clinton was known to make compost and use it on his herb garden when he was governor of Arkansas.

THE SCIENCE OF MAKING COMPOST

Composting is essentially the process of "cooking" the raw ingredients of yard waste, kitchen scraps, and other organic debris until they are "done." The cooking is really the process of digesting done by millions of microorganisms, fungi, worms, and other bacteria that live in the pile of organic matter.

For those tiny creatures to do their work, they need an ambient air temperature above 40 degrees F., food to eat that is full of both carbon and nitrogen (leaves and grass clippings, for instance), moisture, and oxygen. In cooler climates, when the temperature drops below 40 degrees in winter, microbial activity in the bin stops, only to resume when the temperature warms up again in spring.

Moisture in your bin is critical. The contents should be neither too wet, which means dripping, or too dry. Composting matter should resemble a wrung-out sponge.

Oxygen is also critical. If oxygen is insufficient, the aerobic microorganisms will die or vacate, to be replaced by anaerobic bacteria, which create the proverbial "big smelly mess" in your compost bin. You supply the oxygen by turning the contents occasionally.

Microorganisms are omnipresent in the natural world. You don't need to add any additional microorganisms in the form of activators if you don't want to, although they are quite helpful and therefore recommended. If you have filled the bin correctly and the other conditions are favorable, the microorganisms will take over and start the process.

The heat that is generated in your bin is essentially the heat produced by the metabolic activity of the microorganisms as they oxidize or digest the materials in the bin. Heat is not an essential part of composting, per se; it is an indicator that the aerobic microorganisms are present and active.

A typical backyard compost bin's contents will normally reach not much higher than 104 degrees F. Once it has reached this plateau and begins to cool down, it means first that the contents need to be aerated by stirring, and eventually that the microorganisms, fungi, worms, and other tiny creatures have done their work and compost has resulted.

WHY BOTHER?

Making compost happen does take a little time and forethought, and it does involve a little expense to buy a bin and a turning fork. So why bother? Why should you make compost, anyway?

It's good for the environment. Grass clippings, leaves, other yard waste, and compostable kitchen scraps make up 25 percent of all the trash the average family creates in a normal year. A quarter of all the trash you create can be composted. More than thirty-three states and dozens of counties and towns have already barred the disposal of yard waste, leaving you with the burden in many cases. If you recycle cans, bottles, and newspapers, composting is the next natural step for you to take to be more involved in environmental cleanup.

It's good for the garden. Compost is made up almost entirely of decomposed organic matter, a material your garden, your shrubs, your flowers, even your lawn desperately need to be lush, green, and beautiful. Adding compost to the garden or landscape makes the soil more crumbly and easier to dig, it acts as a sponge to conserve water, and it can control weeds when used as a mulch.

It's satisfying. Making compost is a good way to get back in touch with the natural world in this era of two-hour commutes, too much television, and too little time. You're an alchemist. You turn ordinary leaves, banana peels, and leftover soup into compost.

Everything You Know About Compost is Wrong

Well, not everything, but almost everything. When composting came back into vogue riding the tails of Earth Day, a lot of mainstream magazines and television programs asked their junior editors, most of whom had never made compost in their life, many of whom had no idea what compost was, to prepare stories about making compost. The result was a lot of conflicting information that confused people and resulted in a lot of bad compost making. Let's try to set the record straight:

Crumbly brown compost does not automatically pour out the bottom of a compost bin. You may dig the material out of the bottom of the bin several months after you start your compost pile, after the compost has happened, but it is not as easy as putting things in the top and having compost come right out the bottom.

You can't dump fresh grass clippings in a bin, sprinkle with "fairy dust," and expect compost to happen. Fresh grass clippings tend to mat down and give off an ammonia-like odor when they decay. They have to be mixed with leaves or other high-carbon materials—as well as compost activator—in order to make compost.

Properly made compost does not smell or attract flies. Meat scraps, bones, grease, and other animal products are what make compost smell. Leave them out. Fresh vegetable scraps can attract flies. Cover each addition of kitchen scraps with shredded leaves, grass clippings, or kitty litter.

Compost can be made in as little as two weeks, but don't count on it. The average person will not make compost in two weeks. Give your compost at least six months to happen unless you are using the Super-Intensive Method I describe later. Be patient.

You don't have to truck fresh manure in from Old MacDonald's farm to make compost. I use horse manure in my compost bin because I live in the country and a horse lives next door. Manure is a wonderful addition to any compost bin, but it isn't practical for most people. Just make compost out of what you have available, and you'll be all right.

You don't have to turn the compost pile every week. No way. Turn or stir your compost once a month and forget about it.

WHAT CAN AND CANNOT BE COMPOSTED

The simple rule for what can and cannot be compost is this: If it comes from a plant, toss it in; if it comes from an animal, toss it out. Finished compost will take at least six months and will resemble crumbly brown soil—perfect for use in planters or for adding to the lawn or garden. Compost helps loosen the soil, helps attract beneficial worms and microorganisms, and helps make your lawn or landscape more drought resistant and the most beautiful it has ever been.

This is the most important information in this chapter. What you put in or don't put in your compost bin will cause you more success or headaches than any other element of composting. Before I give you the big list, here are a few simple tips:

1. Slice, chop or shred everything as fine as possible. Large chunks of broccoli stalks, whole grapefruit rind, and sticks take longer to compost than finely chopped pieces.

2. Good compost takes a mixture of materials high in carbon, such as leaves and paper egg cartons, and nitrogen, such as grass clippings and vegetable scraps. Add more carbon than nitrogen materials and read the "Science of Composting" section in this chapter.

3. If it comes from a plant, compost it. If it comes from an animal, throw it out, except for manure.

YES

ashes (from wood, not charcoal))	cat litter
eggshells, crushed	peanut shells
vacuum cleaner contents	sawdust
wood chips, bark	tea bags, tea leaves
coffee grounds and filters	hair
paper egg cartons	molded fiber paper plates
paper napkins	paper towels
shredded Xmas trees	shredded paper bags
dead flowers	hay and straw
fruit and vegetable peels	fruits and vegetables
grass clippings	shredded leaves
shredded hedge trimmings	corncobs
garden plants	house plants
leftover salad	barnyard manures
pasta	bread

NO

animal fat	meat scraps
chicken fat and bones	fish and bones
cheese	grease and oil
barbecue ashes	clothing
rocks	large pieces of wood
shellfish shells	circulars and junk mail
pet feces	

THREE METHODS

There are three different approaches to making compost. There are the **Just-Make-a-Big-Pile Method,** favored by old hippies, unrepentant back-to-the-landers, and great-grandparents who like to do things the old-fashioned way; the **Super-Intensive Method,** favored by retired military officers, high school science teachers, and people who either have too much time on their hands or are obsessed with compost making; and the **Goof-Proof Method,** favored by ordinary people with busy lives who want to make compost that doesn't smell and actually produces a usable product that can make their gardens look better.

I or my friends have made compost using all three of these methods. The knowledge you are about to receive is based on firsthand experience, trial and error, and research. Good luck.

The Just-Make-a-Big-Pile Method

This is the low-tech, low-cost, easiest-to-assemble but slowest-to-happen method. You simply make a large pile of leaves, grass clippings, kitchen scraps, barnyard waste, or whatever, and that's it. You can surround your pile with chicken wire or storm fence or cover it with a tarp, but you don't have to. Rain and/or snow will provide all the moisture you need.

This method is for the rural or semi-rural gardener, because a good wind can blow your compost pile all over the yard, and it is easily infiltrated by critters of all shapes and sizes.

Also, give this pile a good one to two years to decompose. It'll take a long time, but, hey, who cares?

The Super-Intensive Method

This is the most scientific way of making compost. It takes the most time to formulate and build the pile but the least time to actually make compost. You get compost that is high in nutrients and usable organic matter that will really energize your soil. If you've got the time to devote to making it right, this is the way to go.

1. Place your bin or tumbler on level ground anywhere that is convenient for you. Full sun or semi-shade is best.

2. Shred all the grass clippings, leaves, spent garden debris, hedge clippings, etc., into the smallest chips possible using a power shredder.

3. Fill the bin or tumbler with all the materials you have, stirring each new addition into the previous matter and sprinkling a cup of compost activator over the material each time.

4. When the bin is one-third full, slowly pour 1 gallon of water over the pile. When the bin is two-thirds full, add another gallon, and add yet another gallon when it is full.

5. When the bin is full, let the materials rest for two to three hours or overnight. The pile will shrink as the weight of the materials pushes a lot of the air out.

6. Add more materials to the pile, layering, stirring, sprinkling, and watering one more time, until the bin is full again.

7. Let the compost rest for one week to give the microorganisms time to start their digestive action. Plunge a compost thermometer into the center of the pile for a reading. You should get a reading of over 120 degrees F. Check the pile's temperature daily until it begins to lower. Higher than ambient temperature means the microorganisms are metabolizing the raw materials. When the temperature cools, it means the microorganisms have consumed the available raw ingredients and need new ones, which are supplied by turning.

8. Using a turning fork or an aerator tool, turn the materials in the compost pile, stirring them completely to incorporate the cooler outer portions with the hotter inner portions. Add a gallon of water and wait another week. Test the temperature again.

9. Continue this temperature testing and turning every week. You should start noticing that the compost is beginning to turn brown and resemble coarse soil.

10. Let the compost cool slightly and then spread it around your garden as necessary.

The Goof-Proof Method

Most people don't want a big, unsightly pile of compost in their yard, nor do they want to concentrate so hard on making compost. Most people want to set up an inconspicuous compost bin, fill it full of yard waste and kitchen scraps, and let compost happen.

Goof-Proof Compost is geared to a regular person's life. You fill the bin with what you have—leaves and grass clippings—add kitchen scraps every day or two, stir it up once in a while, and eventually you'll have compost. It might take six to eight months, especially in the winter, but, at least you've found a good place for all that yard debris and those kitchen scraps and you haven't spent too much time or energy thinking about compost.

First, buy a hard-plastic, wooden, or heavy wire bin with a lid. If you compost kitchen scraps, you must have a lid to keep the critters out. Place the bin in a semi-shady place or wherever the best place is in your yard. Be sure the bin has holes or slats for air to circulate.

Then, just keep adding leaves, grass clippings, kitchen scraps, dead plants, etc., to the pile, cover up the fresh food scraps every time, and forget about it. As long as you cover the food scraps, you won't have flies or odors. The more shredded or broken up the materials are, the faster the compost will happen.

Never, ever, add meat; chicken; fish; their bones or their grease; cheese; clam, oyster, or mussel shells—or grapefruit rinds that have not been chopped up. Every year I get lazy and toss a few chicken bones in the compost, and every year I find them intact when I empty the compost bin.

Keep a spare bag of shredded leaves, broken-up pine needles, a bale of hay or straw, or even a bag of kitty litter to sprinkle over the kitchen scraps every time you add them. Rain and/or snow plus moist kitchen scraps will usually provide all the moisture you need. If you find the pile to be dry, pour a couple gallons of water on it when you turn it.

Once a month turn or stir the pile so that no clumps are forming that might smother the pile and prevent the compost from happening. If you want to make your compost happen a little faster, add a cup of compost activator or a shovelful of garden soil to the pile when you are stirring. Activator and soil are full of microorganisms and nutrients that speed the process along.

In a few months, you'll notice that the compost is starting to look more like soil than garbage. Turn the contents one more time, cover it, and stop adding any more materials. Wait one month and your compost will be ready to use. In the meantime, just save all your debris and scraps to reload the bin once it is empty.

HOW DO I KNOW WHEN MY COMPOST IS "DONE"?

Your compost is done when all the yard debris, kitchen scraps, and everything else you have put into the bin look like crumbly, earthy-smelling soil. It is not soil yet, but it looks a lot like it.

The real value of compost is that it provides an excellent source of organic matter, which along with sand, clay, rocks, microorganisms, air, and moisture make up what we call soil.

Plants growing in a soil rich in organic matter are stronger, healthier, less susceptible to pests and diseases, and less prone to suffer from drought; they produce fruits, flowers, and leafy growth to their full potential. In other words, compost will give you the best-looking, most productive garden, landscape, and lawn around.

Shovel the finished compost into a wheelbarrow, garden cart, bushel basket, or bucket. With a shovel or a garden trowel, spread the compost around and between annual or perennial flowers, shrubbery, herbs, fruits, and vegetables, at the base or drip line of trees, and over your lawn.

Compost is great for your lawn. Your lawn needs organic matter added each year for the same reasons as your garden. Simply scatter the compost over your lawn to a depth of no more than ½ inch. Then give it a light raking to break up any small clumps that might kill the grass and to make it disappear down to the soil line. The best time to add compost to your lawn is when you have the compost and feel like doing the work.

Top Ten Ways to Clean, Easy Compost

1. Forget everything you've read or heard about compost up to this point. Most of it is wrong, at worst, or confusing, at best.

2. Buy the right bin. Get a hard-plastic, wooden, or sturdy wire bin with a lid, that measures approximately three to four feet square. If you compost kitchen scraps, your bins should have a lid.

3. Place it in the right spot. Look for a level, semi-shady or sunny spot that is close enough to the house so you will actually use the darn thing.

4. Do fill it with: grass clippings, shredded leaves, fruit and vegetables and their peelings, coffee grounds and filters, paper egg cartons, paper napkins and towels, pasta, rice, beans, hedge trimmings, kitty litter, dead flowers and plants, fireplace ashes.

5. Don't fill it with: pet poop, meat, chicken or fish flesh or bones, grease or oil, cheese, barbecue ashes, junk mail.

6. Shred everything if possible. The smaller the pieces that go into the bin, the easier it is for the microorganisms inside to turn them into compost.

7. Add compost activator and water. Activator gives the compost a "jump start," and water provides the necessary moisture to keep the pile going. You can also add worms, garden soil, or manure to help make the compost happen.

8. Cover each new addition. Every time you add kitchen scraps to the bin, cover them up with leaves, grass clippings, kitty litter, or soil to prevent flies and odors.

9. Put a lid on it. If you are composting kitchen scraps, the only way to keep neighborhood cats, dogs, raccoons, coyotes, deer, and other critters out of the bin is with a sturdy, tight-fitting lid. Do it.

10. Stop worrying. Be patient. Relax. It might take your compost six to eight months to happen, but it will.

Choosing the Right Compost Bin—
An Opinionated Survey

WIRE COMPOST BIN

A.k.a. E-Z Compost Bin, Compost Tamer. Made of vinyl-coated steel wire, 36 inches square x 30 inches high. About $40.

Appearance: The green vinyl coating on the wire spruces up this simple bin, which resembles a child's playpen or an old-fashioned yard fence.

Attributes: Easy to assemble, about five minutes, with four panels and four slip-in corner rods. Low price makes this an ideal bin for the first-time composter. Easy access for turning and aerating. The 2-inch-wide openings in the fence-like sides give the compost good aeration but could be a tempting target for varmints.

Availability: Gardener's Supply Catalogue, 128 Intervale Road, Burlington, VT 05401, (802) 863–1700; Natural Gardening Co., 217 San Anselmo Avenue, San Anselmo, CA 94960, (415) 456–5060; as well as Ace Hardware and some other retail outlets.

Manufacturer: Northern States Industries, 1200 Mendelssohn Avenue, Suite 210, Minneapolis, MN 55427, (612) 541–9026.

WESTERN CEDAR BIO BIN

Made of recycled western cedar, 35 x 35 x 31 inches high. Holds fourteen bushels of compost. About $60.

Appearance: Has a natural elegant look of weathered western cedar. It is rot resistant and will last for many years. Made of recycled scrap lumber from second-growth trees. This is an ideal bin at a good price for people who like the look of wood.

Attributes: Easy to assemble, in about ten minutes, by slipping cedar slats over four steel rods. Slatted sides give lots of aeration. Easy to turn and aerate materials. You can buy a lid for the bin, also made out of recycled cedar, for $15, and you can buy add-on modules to double the size of your bin.

Availability: W. Atlee Burpee & Co. Catalogue, 300 Park Avenue, Warminster, PA 18974, (800) 888–1447.

Manufacturer: Heritage Products, 8407 Lightmoor Court, Bainbridge Island, WA 98110, (206) 842–6641.

Western cedar bio bin and wire compost bin.

EASTERN CEDAR COMPOSTER

Made in Maine of eastern cedar, 38 x 41 x 32 inches high. Holds twenty-six cubic feet of compost. About $85.

Appearance: Sturdy, good-looking wooden compost bin with slanted horizontal slats. Natural golden wood finish weathers to an attractive silvery gray resembling shingles on a Cape Cod house. Good for backyard or more rural setting.

Attributes: Easy to assemble, about ten minutes, with pre-cut wooden panels, but you will need a screwdriver. Front panels slip out to make turning and unloading easy. Rot-resistant cedar will last for a very long time.

Availability: Johnny's Selected Seeds Catalogue, Foss Hill Road, Albion, ME 04910-9731, (207) 437–4301; L.L. Bean, Freeport, ME; Eddie Bauer, P.O. Box 3700, Seattle, WA 98124, (800) 426–8020; and many Ace and True Value hardware stores.

Manufacturer: K-D Wood Products, Lander Avenue, Bingham, ME O4920, 207–672–4333.

THE SOIL SAVER

A.k.a. the Cadillac of Composters. Made of heavy-duty, 50 percent recycled, post-consumer plastic, 28 x 28 x 30 inches high. Holds twelve cubic feet of compost. $60 to $100.

Appearance: Square and compact, made from hard black plastic, it resembles a central air-conditioning unit. Good-looking compost bin especially recommended for urban and suburban yards, because you never see the yard and kitchen waste once you get it inside.

Attributes: Guaranteed to last ten years. This is an excellent bin, very sturdy, with a tight-fitting lid. Highly recommended. Has sliding bottom doors and fourteen nylon carriage bolts, which makes assembly a little more difficult. It takes about thirty minutes to assemble using a special tool supplied in the kit.

Availability: True Value, Ace, Home Depot, Hechinger's, Frank's Nurseries, Aubuchon Hardware, ServiceStar, and many other retail stores. Mail order from Walter Nicke, 36 Mcleod Avenue, Topsfield, MA 01983, (508) 887–3388; Jackson & Perkins, 518 South Pacific Highway, Medford, OR, 97501, (503) 770–2675.

Manufacturer: Barclay Recycling, Inc., 75 Ingram Drive, Toronto, Ontario M6M 2M2, Canada, (416) 240–8227.

THE HUMUS BUILDER

The junior size of Soil Saver. Made of 50 percent post-consumer polypropylene plastic, 26 x 26 x 30 inches high. Holds 9 ½ cubic feet of compost. $35 to $50.

Appearance: Black, hard-plastic shell; this is a more streamlined version of the Soil Saver. Its smaller, more compact look makes it perfect for smaller backyards or even rooftops and balconies for city gardeners.

Attributes: Guaranteed for one year. Very easy to put together, ten minutes, snaps together with no bolts or tools required. Has a snug-fitting lid; it's lightweight and easy to move around. Good ventilation for aeration.

Availability: Gardener's Supply Catalogue and most of the same retail outlets and direct mail catalogues as the Soil Saver.

Manufacturer: Barclay Recycling, Inc., 75 Ingram Drive, Toronto, Ontario M6M 2M2, Canada, (416) 240–8227.

RUBBERMAID COMPOSTER

Made of heavy-duty green plastic, 36 x 33 x 27 inches high, rectangular shape. Holds eighteen cubic feet of compost. About $75.

Appearance: Forest green, hard-molded plastic. It has a more rectangular shape than most compost bins and a slanted roof that resembles a Dutch barn. On the whole, this composter looks a lot like a dog house.

Attributes: The lift-off roof makes this compost bin very easy to load and to turn. It is incredibly easy to assemble, about five minutes, with the large panels popping together on large slotted hinges. This makes it easy to load and unload and to stir with a fork or aerator. This composter is also highly recommended for suburban or urban yards. Virtually critter proof.

Availability: Widely available at most large hardware stores, home centers, some supermarkets and discount department stores. No catalogues.

Manufacturer: Rubbermaid Incorporated, Specialty Products Division, 1147 Akron Road, Wooster, OH 44691.

RINGER COMPOST KING

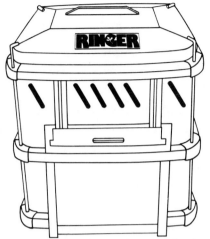

Forest green, hard-plastic square construction, 33 x 33 x 31 inches high. Holds twelve bushels of compost. About $120.

Appearance: This dark green compact bin with a tight-fitting lid makes it ideal for a suburban or urban backyard. It is a little bit pricey at nearly $120, but it is very well constructed and will last for many years.

Attributes: The thick walls of this bin hold moisture and heat very well, which provides a nice warm atmosphere for microorganisms during the cooler late fall and early spring months, when composting would otherwise slow down. This bin is fairly easy to assemble, about fifteen minutes, with no tools required.

Availability: W. Atlee Burpee & Co., some retail outlets, and the Ringer Catalogue, (800) 654–1047.

Manufacturer: Ringer, 9959 Valley View Road, Eden Prairie, MN 55344–3585, (612) 941–4180 or (800) 654–1047.

YARDMASTER STACKING COMPOST BIN

Black, hard plastic, 28 x 28 x 33 1/2 inches high. Holds 12.8 cubic feet of compost. About $100.

Appearance: This is a basic-looking square bin made of hard black plastic. It has a tight-fitting lid, and the whole bin should look nice in an urban or suburban backyard.

Attributes: The YardMaster is the only bin on the market that is stackable. It is comprised of three panels stacked on top of each other, a design that is very handy for turning compost. You simply remove the lid and the first unit, start filling the first unit with compost, remove the second unit, and so on until the compost pile is turned. There is nothing better for making compost than turning it once a month.

Availability: Mail order from Nichols Garden Nursery Catalogue, 1190 North Pacific Highway, Albany, OR 97321, (503) 928–9280; Smith & Hawken, 25 Corte Madera, Mill Valley, CA 94941, (415) 383–2000.

Manufacturer: Southern Case, Inc., 2315 Laurelbrook Street, Raleigh, NC 27604, (919) 821–0877.

Compost Activator—Good Idea or "Snake Oil"?

Compost activator is a brown, black, or green powdery substance that you sprinkle on your compost pile to help make compost. There must be at least two dozen different brands of compost activator on the market today. The manufacturers don't tell you what's in their activator, but I will. The bulk of the product is composed of the same materials used to make natural organic fertilizer: soy, wheat germ or cotton seed meal, blood meal, bone meal, sometimes seaweed, or other agricultural by-products. In addition, some activators have added microorganisms.

You don't absolutely, positively need to use compost activator. The kinds of microorganisms in compost activator are already present in the yard waste and other debris you load into your compost bin. Furthermore, a well-made compost pile already has a good combination of nitrogen and carbohydrates, the foods the microorganisms like to eat as they make your compost. If you are patient, your compost will happen without the use of activator.

Having said all that, I still suggest that you use some sort of compost activator. Why? Because maybe the microorganisms in your compost bin are lazy, or you don't have a good blend of nitrogen and carbon. Adding compost activators to your compost bin is like jump-starting your car. They give your compost a little boost to get it going and hopefully keep it going until it's done.

Low-cost alternatives to compost activator include any type of natural organic fertilizer you use for your lawn and garden, fresh manure of any kind, a few shovels full of good soil from your garden, or leftover compost from an existing bin.

Compost Tools—Aerators and Forks

If you want to make compost in a reasonable amount of time, you must stir up the contents periodically. First, the microorganisms at work in your bin need fresh air, oxygen, to breathe. Stirring the compost brings fresh air into all parts of the pile. This is called aeration.

Next, compost happens at the center of the pile of yard waste and debris. It is

You need to turn the contents of your compost bin about once a month to mix the rapidly decaying matter in the center of the pile with the raw ingredients on the outside of the pile. Pitchforks and forked spades are fine, but try using a compost aerator tool like this one to cut down on the back-breaking labor. Plunge the tool into the pile, twist it, and pull it out. It does the mixing for you.

important that you stir the cooler outer realms of debris into the hotter inner circle. This is called turning.

A compost aerator is a specially designed tool that looks like a short pogo stick with a tiny propeller on one end. Here's how it works: Hold the handles with both hands and plunge the blade end down into the compost. As you pull the tool back out of the compost, the blades open up, turning it and aerating it at the same time.

There are several aerators on the market, but the one I recommend is the E-Z Turn Compost Tool by Dalen Products, 11110 Gilbert Drive, Knoxville, TN 37932. The E-Z Turn is a heavy-duty model, made of galvanized steel, and offers a full lifetime guarantee. At 34 inches, it's longer than most tools, and it has a double handle with rubber grips, two features that make it much easier to use and actually do a better job of turning and aerating the compost. Price $15; widely available at home centers, lawn-and-garden shops, and hardware stores.

Compost forks and forked spades are two other tools you can use to turn and aerate compost. Both of these resemble small pitchforks, and they are usually made with D-shaped handles, which makes them easier to use. Neither one, however, does a good job aerating the compost while it is still in the bin. Using these to aerate a pile is simply a back breaker. I mention them here in case you already own one or both. On the whole you're better off with an aerator and a shovel.

But, once you have finished composting, they are useful for unloading the compost from the bin. If you already have a shovel or a garden spade, you can use that tool to unload the bin as well.

Three Success Stories

SEATTLE. The Seattle Tilth Association is a large group of urban gardeners who have been interested in composting and soil saving for many years. In 1986 they started working with the Seattle Department of Solid Waste to tackle the city's garbage crisis. They set up five demonstration sites, complete with compost bins and information, to show people how they could compost their yard waste to save the environment. Since then, they have given out nearly twenty thousand compost bins, many made of recycled milk jugs, to citizens, to encourage people to start composting at home. For information, contact the Seattle Tilth Association, 4649 Sunnyside Avenue North, Seattle, WA 98103, (206) 633–0451.

ST. LOUIS. The Missouri Botanical Garden, one of the world's most prestigious horticultural institutions, has gotten into composting in a big way. They sell compost bins in their gift shop; they have produced a home video on composting; and in 1991 they started a Master Composter Program.

Since then, they have trained more than one hundred Master Composters, equipped them with slide shows, and sent them out to civic groups such as the Lions Club and major banks to get them interested in the virtues of composting. The state of Missouri banned the disposal of yard waste in 1992, making composting a virtual necessity. For more information, contact the Missouri Botanical Garden, P.O. Box 299, St. Louis, MO 63116, (314) 577–9561.

NEW YORK CITY. Next time you see the famed horse carriages parading around Manhattan's Central Park, remember that those horses are part of a lean, mean composting machine right in the heart of Gotham. All of the manure those horses create at their stables is being made into compost in bins near the Hudson River docks. A group called the Green Guerrillas is working with several local garden clubs to compost stable waste and kitchen waste from people's apartments. The finished compost is then used to spread at the base of city trees and bushes in parks and other places where it can do some good. The manure from Central Park's riding horses and the police department's horse stables is also being composted, as well as thousands of residents' annual Christmas trees.

For more information, contact the Green Guerrillas, 625 Broadway, 2nd Floor, New York, NY 10012, (212) 674–8124.

Part Two

REGIONAL

LAWN CARE GUIDE

L awns are getting to be a lot more like gardens these days. It seems that there is almost always something we can do to or for our lawn, even in the cold of winter. And what we do depends on where we live. People living in southern regions, for example, have no cold of winter. Their grass can grow all year round. Lawn care for them is very different from lawn care for people living in, say, the Northeast.

A lot of us are actually rank beginners when it comes to lawn care. You may be moving into a new house, where you'll be faced with your first lawn, or you may have moved to a different part of the country where lawn care is on a different schedule. There are a lot of reasons why we might not know exactly what to do to our lawn and when to do it.

This next section—actually four regional lawn care guides—gives a month-by-month schedule of activities you can and should be doing to have a good-looking and healthy lawn. Lawn care is changing rapidly with the introduction of new fertilizers, weed and pest controls, and even new ways to mow and water. What I have tried to do in this section is give you a schedule for handling this new age of lawn care. For instance, I suggest that you control grubs with beneficial nematodes rather than with chemical grub control. The latter is usually applied in spring and late summer, when grubs are active. Beneficial nematodes, however, live in the ground all season long and are best applied only in spring.

In addition to pest control, you will also find monthly cues on watering, planting, mowing, maintaining your equipment, controlling weeds, and more.

The United States and Canada can easily be divided into two great grass-growing regions. Cool-season grasses—grasses that prefer to grow in spring, early summer, and fall—grow best in the northern half of the United States and most of Canada. Kentucky blue, fine fescue, and perennial rye are cool-season grasses. Warm-season grasses—grasses that like to grow and stay green during the hot summer months—grow better in the southern United States. Bermuda, Bahia, zoysia, and St. Augustine are warm-season grasses.

But there are so many regional variations with altitude and people's willingness to irrigate their lawns, that I've found it better to break the country up into four regions: the cool and humid Northeast, Midwest, and Pacific Northwest; the hot and humid South; the hot and dry Southwest; and the high and dry Great Plains.

I encourage you to read the first four chapters of this book, before consulting the regional guide because they will give you a full understanding of environmental lawn care as well as the specific do's and don'ts of mowing, watering, and grass selection—even the right kind of mower to use.

You'll see that I've included composting in the monthly activities. Why composting? Because I think that composting is an integral part of lawn care. The leaves that fall on your lawn, some of the grass that you mow, hedge trimmings, spent flowers, and other lawn debris all make excellent compost. You can use that homemade compost as mulch around your flower beds and under your shrubs. You can also use it to enrich the soil when you plant new grass.

You don't have to have a garden to make compost. Most yards have some sort of tree or shrub that drops leaves or needles. If you have a yard, you can make compost. Composting is the environmental way to manage your lawn debris. If you haven't started making compost yet, please read the "Clean, Easy Compost" chapter in this book for complete instructions.

As you read through the suggested activities in the different months, you will notice that not all months contain instructions for each and every category. If there is nothing significant to do in a particular category in a given month, I have omitted the category from that month's instructions.

Finally, please take advantage of the knowledge your local lawn-and-garden center and your Cooperative Extension office can give you about what works best in your area.

The Cool and Humid Northeast, Midwest, and Pacific Northwest

~

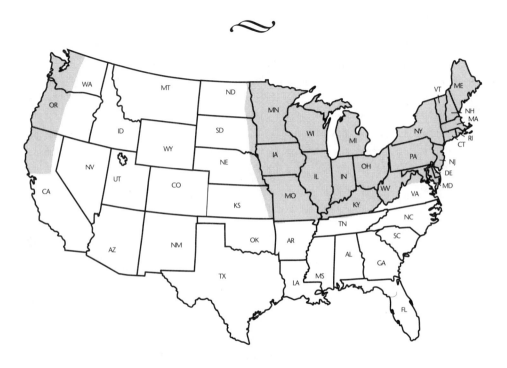

This area of the country runs from Maine to Maryland, west to Topeka, Kansas, and Omaha, Nebraska, north to southern Ontario, and back to Maine. It includes all of New England, New York, New Jersey, Pennsylvania, West Virginia, Kentucky, and the Midwest. It also includes the Pacific Northwest west of the Cascades, Seattle and Portland, and as far south as parts of northern California.

The soil is generally acid, although there can be significant local variations, and is a mixture of sand and clay. The summers are warm, the winters are cool to cold, and there is ample rainfall to grow the cool-season grasses such as Kentucky bluegrass, perennial rye, and fine fescue.

JANUARY

WHAT TO WATCH FOR

By now, the really cold weather, complete with snow and freezing rain, has started to take hold. If you're lucky, you will get a blanket of snow to cover your lawn all winter, which acts as a protective barrier for your lawn from the damaging effects of freezing winds.

Watch out for salts and other de-icing chemicals that the snow-removal road crew may be putting on the roads and then plowing on to your lawn. Those chemicals can damage your grass. You can avoid damaging the grass along your own walkways and driveways by using sand, kitty litter, or sawdust along your walks instead of rock salt. You might even consider tossing out a granular natural organic fertilizer that will wash off into the lawn and give it a little boost of nourishment come spring.

MOWING AND MAINTENANCE

Since you can't mow this month, you should take your lawn mower blades in for sharpening. If you have an electric mower that prefers continuous charging, be sure that it is still plugged in.

FERTILIZING AND SOIL BUILDING

It's not too early to start poking around your lawn-and-garden center asking if they are going to have a good supply of natural organic fertilizer, lime, soil-test kits, and whatever else you are going to need to build your soil once the season begins.

If you don't have a compost bin going already, please, get one started. Read the "Clean, Easy Compost" chapter in this book.

Shred your Christmas tree and add it to the compost bin along with kitchen scraps.

WEED CONTROL

If you are having a mild winter, you might be able to go out and dig up dandelions or plantains in your yard.

PLANTING AND OVERSEEDING

You can't plant yet, but you can start shopping for the right grass seed for your area. Although the blends of endophytic perennial rye, fine fescue, and disease-resistant Kentucky bluegrass are widely available, you might want to make sure your lawn-and-garden center will be carrying what you want later in the spring.

WATERING

Snow and rain should be supplying lots of water this month. But how much? Either set out your own rain gauge or keep track of the precipitation as it is reported in your local newspaper or from the TV meteorologist. Light amounts in the winter could lead to drought conditions in the summer.

FEBRUARY

WHAT TO WATCH FOR

Most of the northern and central parts of this region will still be buried under snow, frozen ground, and cold weather. Some of the southern-tier counties in Kentucky, Maryland, and Missouri may get an early thaw and begin to see some green grass shooting up.

MOWING AND MAINTENANCE

Now is a good time to take your mower in for a tune-up, before everybody else decides to do it and the repair shop gets overcrowded.

Now is also a good time to shop for a mower, when the selection is great and dealers are eager to do business.

FERTILIZING AND SOIL BUILDING

Continue to make compost with kitchen scraps, even if the temperature is below 40 degrees and the material in the bin is frozen. Gather and shred any fallen tree limbs or leaves left over from last fall and add them to the compost pile. If you get a few warm days, try to turn the pile to get it going again.

WEED CONTROL

You can't do much to control weeds this month except digging out any dandelions or plantains that are beginning to appear.

PLANTING AND OVERSEEDING

It is still too early to start planting, but you should have your seeds firmly in hand because you might be able to plant next month.

PEST CONTROL

The only thing you can do about pest control this month is to ask your local lawn-and-garden center if it will have grub-busting beneficial nematodes that you can buy later in the spring. As the soil begins to warm, grubs will be migrating toward the surface.

WATERING

Continue keeping track of your precipitation so that you will know whether the winter rains and snow have been providing average amounts of moisture for your soil.

WHAT TO WATCH FOR

March is a wild and wacky month weatherwise—you never know what to expect. Some areas will still have snow, followed by mud, but others will have grass growing and the season fully under way. Get ready!

MOWING AND MAINTENANCE

Be sure your mower is tuned up or charged up, the blades sharpened, the oil changed, and fresh gasoline put in. Before you start mowing, be sure to give your lawn a good raking to gather up any dead limbs, rocks, leaves, and other debris that can damage the grass and provide weeds with an opportunity to sprout.

Set the cutting height to $2\frac{1}{2}$ to 3 inches, and start mowing when the grass is ready to be mowed.

FERTILIZING AND SOIL BUILDING

Spring is the second best time to apply fertilizer, but wait until next month, after the grass has gotten its first good green growth spurt.

Now is a good time to take a soil test. Simply take your soil samples to your Cooperative Extension office or lawn-and-garden center, or do it yourself. If you need to add lime, which is likely in this region, now is a good time to apply it, either with a spreader or by broadcasting it manually.

Now is also a good time to top-dress your lawn by sprinkling on and raking in a $\frac{1}{4}$- to $\frac{1}{2}$-inch layer of sifted compost, dehydrated manure, or peat moss. Top-dressing adds valuable organic matter to the lawn.

WEED CONTROL

Setting your mower height to $2\frac{1}{2}$ to 3 inches is the best way to prevent the emergence of annual weeds such as crabgrass.

Dig out perennial weeds such as dandelion and plantain, and pull out knotweed, chickweed, and ground ivy as they emerge.

PLANTING AND OVERSEEDING

Now is the second best time to plant cool-season grass seeds in the southern parts of this region. You can begin planting once air temperatures are consistently in the 70- to 75-degree range. Always plant a mixture of endophytic perennial ryegrass and fine fescue plus a little disease-resistant Kentucky blue that is blended for either sun or shade, depending on your conditions.

PEST CONTROL

Now is a good time to apply grub-busting beneficial nematodes if you feel your lawn is being damaged by grubs. Be sure to wait until the ground is thawed and there is no longer any chance of frost in your area. Then follow the manufacturer's instructions *very* carefully.

WATERING

Do not water your lawn at this time. You should be getting adequate rainfall.

WHAT TO WATCH FOR

This is a beautiful month for the lawn. The crocus, daffodils, and tulips are blooming in many places. The weather is cool and rainfall ample, which is exactly what your cool-season lawn loves.

Watch out for bare spots in the lawn that might need patching and repair them.

MOWING AND MAINTENANCE

You are mowing a lot now, maybe as much as every four days. Remember to follow the one-third rule (page 16) and never cut off more than one-third of your grass's height at any one time. If you have a side-discharge mower and you feel the clippings on the lawn are too dense, rake them up and put them in the compost bin.

FERTILIZING AND SOIL BUILDING

Now is the time that most people in this area will be ready to apply natural organic fertilizer. Fall is really the better time, but spring is all right, too. In general, a 2,500-square-foot lawn will need about 25 pounds of natural organic fertilizer each year. Apply half of it now and the rest in the fall.

WEED CONTROL

Keep the mower set high and dig out dandelions, plantains, and other perennial weeds. As you dig out the weeds, be sure to spot-patch the areas with new grass seed to let the grass muscle the weeds out of the lawn.

PLANTING AND OVERSEEDING

People living in the northern areas of this zone can start planting grass seed now. Be sure to prepare the soil, add natural organic fertilizer, sprinkle on plenty of grass seed, and keep it watered until it germinates.

PEST CONTROL

As grubs migrate toward the soil surface, now is another good time to apply beneficial nematodes according to manufacturer's instructions.

WATERING

Unless you are planting grass seed, don't water. It is still too early. If you start watering now, your grass will develop shallow roots that will make it drought prone later in the season. Watering now could also contribute to the spread of lawn diseases.

WHAT TO WATCH FOR

Watch to make sure your grass is growing strong and green this month. If it looks weak and weeds are getting a hold, you need a soil test and may need more fertilizer, top dressing, and other help such as new grass seeds later in the season to correct the problems.

MOWING AND MAINTENANCE

Keep mowing. If your lawn is large, you may want to consider having your blades sharpened again. Dull blades beat, tear, and shred the grass, causing damage.

FERTILIZING AND SOIL BUILDING

Finish up your fertilizer applications. You are building the soil now by mulching your grass clippings into the turf, where they will decompose and add nitrogen fertilizer and organic matter to the soil.

WEED CONTROL

Now is one of the best times to dig dandelions, which they are blooming and before the seed pods start to blow in the wind. Dandelions have just used up all their food reserves to create that flower, and the plant is at its weakest and most susceptible to digging.

PLANTING AND OVERSEEDING

It is getting late for any serious seeding now because of the emergence of annual weeds such as crabgrass. Hold off until late August to start seeding again.

PEST CONTROL

This is the month that chinch bugs begin making their appearance and start chewing the tops of your grass plants. If you are having a bad infestation, you will want to replant your lawn this fall with the perennial ryegrass and fine fescue that have been treated with endophytes that repel chinch bugs.

WATERING

It is still too early to water unless you live in some of the southern zones of this region. Your lawn will naturally go dormant during the hot dry summer months and return to green when the fall rains return in September. If you really want to water, water no more than once a week and then deliver about an inch of water to the lawn so that it seeps down to a depth of 6 to 8 inches.

JUNE

WHAT TO WATCH FOR

Now that the trees are fully leafed out and the bushes are or have been in bloom, it's a good time to make a full assessment of your lawn and yard environment. If the grass is weedy and not growing well, you may want to plan for a renovation in late summer. Your trees may have grown to the point that you now have a shady lawn area, which would demand grass seed that does better under those conditions. You may also want to shrink your grass's range by planting some ground covers under trees and along the garage. Now is a good time to plant ground covers such as wild ginger, myrtle, ivy, and others.

MOWING AND MAINTENANCE

You should still be mowing every four or so days to keep up with your environmental lawn-care program as outlined in the first chapter of this book. It will pay off for you, because you won't have to use chemicals.

FERTILIZING AND SOIL BUILDING

Don't add any more fertilizer now. You can add that layer of top dressing though, if you didn't get to it earlier in the season.

WEED CONTROL

Now is a good time to assess the amount of weed infestation you have in your lawn. If you've been following the ten-point program featured in chapter one, and you still have serious weed problems, you probably want to consider replanting your lawn later in the summer, after you have taken a soil test.

PLANTING AND OVERSEEDING

The only planting or overseeding you do now should be restricted to spot patching and emergency planting caused by digging or major landscape changes. Wait until August.

PEST CONTROL

By now your grubs are getting ready to emerge from the ground as Japanese and other types of beetles. They will be as close to the surface as they will be all year. Check for them. Dig up a 1-foot-square patch of turf. Count the grubs. If there are more than eight or ten, you have the potential for grub damage. That doesn't mean that you will have damage, but the conditions are favorable. Other signs of grub infestation are the presence of moles in the yard and frequent sighting of birds digging for grubs.

WATERING

I still suggest you hold off watering your lawn unless you are in a serious drought. If you do water, be sure to water no more than once a week and then for an hour or so.

WHAT TO WATCH FOR

Drought and heat will be your lawn's biggest challenges this month.

MOWING AND MAINTENANCE

The grass will not be growing much this month and may even have gone into a light brown dormancy. Don't worry; as the weather cools off and the rains return, your lawn will reappear.

FERTILIZING AND SOIL BUILDING

Do not add any fertilizer to the lawn this month. Do continue to add kitchen scraps, especially all those good fruit and vegetable peelings you are now accumulating, to the compost pile. Turning the compost pile every two weeks will speed up its process.

WEED CONTROL

Spot check for weeds and remove them as necessary.

PLANTING AND OVERSEEDING

Now is not a good time to plant grass seeds, but August is. Get your seeds and strategy ready and plot out a couple of work days to get the job done right.

PEST CONTROL

Beetles will be laying their eggs, which turn into grubs, now and in August. Avoid light, frequent watering of the lawn, because that will provide the beetles with a perfect egg-laying environment. It is better for the lawn to be dry, which helps dry out the eggs and kill many of them.

WATERING

Watering now can help promote grub growth and the spread of disease. Hold off unless you feel it is really necessary.

WHAT TO WATCH FOR

Watch for your lawn to begin to revitalize itself, especially later in the month as days get shorter and the nights get a little cooler.

MOWING AND MAINTENANCE

Continue your regular light and infrequent summer mowing pattern.

Mid- to late August is a great time to plant grass. If overseeding, lower the mower's cutting height and cut the grass as low to the ground as possible.

FERTILIZING AND SOIL BUILDING

The only fertilizer you should add now is a high-phosphorus natural organic fertilizer if you are planting grass seeds.

WEED CONTROL

Weeds are very weak now, and it is a good time to dig them out and plant grass seeds in their place.

PLANTING AND OVERSEEDING

Mid-August until late September is a great time to plant grass seeds for a new lawn. You want to do this about six weeks before you expect the first frost in your area. Be sure to add organic matter, starter fertilizer, and a generous supply of expensive grass seed. Keep it watered and keep the kids off it, until it germinates.

PEST CONTROL

If you didn't apply any grub-busting beneficial nematodes in the spring, this is a great time. The nematodes will die once the cold weather sets in, but now is the best time to kill the young grubs before they burrow down deep into the soil for the winter.

WATERING

You may have to start watering this month, if your area is not under water restrictions. If you do, water no more than once a week and then for only an hour or two to deeply soak the ground. If you've been watering all summer, keep it up or you could cause damage to the grass.

WHAT TO WATCH FOR

The return of cooler weather and rain should make your lawn look great.

MOWING AND MAINTENANCE

Get that mower back out and keep on cutting. As a few of the first leaves fall, you can feel free to mow them right into the ground as a great source of phosphorus fertilizer and organic matter.

FERTILIZING AND SOIL BUILDING

This is the best time of year to apply natural organic fertilizer. Add an entire year's worth right now or the other half left over from your spring application. The fertilizer now helps build up the grass's roots, making it stronger for winter and ready to grow fast in the spring.

Now is also a good time to add lime if your soil is acid.

This is also a very good time to add a thin layer of top dressing such as compost to your lawn. If you have been making compost all summer long, it is probably finished by now and ready to be used. If you are going to put it on the lawn, be sure to sift it first, so that no big chunks get tossed on the lawn where they might suffocate the grass.

WEED CONTROL

Keep digging those weeds out by hand. You should have most of them out by now.

PLANTING AND OVERSEEDING

This is an excellent time to plant grass seed for a new lawn or overseed an old one. Be sure to keep the new seeds lightly watered until they germinate.

WATERING

There is probably no need to water the lawn now, as fall rains are beginning to return.

WHAT TO WATCH FOR

The leaves will be falling, and you want to get them raked up as soon as possible, because if left on the lawn they will kill the grass.

MOWING AND MAINTENANCE

Use the bagging attachment on your lawn mower to vacuum-shred your leaves. Pour the leaves directly into your compost bin or in your flower beds for mulch and winter protection.

When you are ready to cut your grass for the last time, lower the height of the mower to 1½ to 2 inches. That clips the grass low enough so that it doesn't fall over, smother the turf, and lead to potential disease in the lawn.

Don't rake and bag your leaves and haul them away. Harvest them as a valuable source of free fertilizer, compost makings, and organic mulch. Mow your leaves into a bagging lawn mower and empty them into the compost bin or spread them around the garden as mulch and winter cover for perennial flowers. If you like to rake leaves, be sure to run them through a chipper/shredder before you add them to the compost bin.

FERTILIZING AND SOIL BUILDING

October is a good time to add natural organic fertilizer, top-dress, or add lime if necessary.

WEED CONTROL

Just like your grass growth, weed growth is starting to slow down.

PLANTING AND OVERSEEDING

It's a little late to plant grass seed now unless you live in the more southerly zones of this region.

PEST CONTROL

There is nothing you can do this month.

WATERING

You shouldn't have to water at all now.

NOVEMBER

WHAT TO WATCH FOR

Frost will become common now, and the ground will begin to start freezing. Be sure to gather up all those leaves.

MOWING AND MAINTENANCE

When you are ready to put your mower away for the season, be sure to let the motor run until it is out of gas, and drain the oil. Take your blades in for sharpening to avoid the rush next year. Clean the mower really well, especially the underside, where clumps may have built up that can damage the coating.

FERTILIZING AND SOIL BUILDING

November is a great time to apply natural organic fertilizer. Put it on anytime this month.

WEED CONTROL

You may still be able to dig or pull out weeds.

DECEMBER

WHAT TO WATCH FOR

There's nothing much going on in your lawn at this time of year. Happy Holidays!

MOWING AND MAINTENANCE

Put your lawn mower away and don't think about it until next year.

FERTILIZING AND SOIL BUILDING

Keep adding kitchen scraps to your compost bin. Shred your Christmas tree and add it to the bin. Turn the bin one more time before the really cold weather sets in. If you don't want to shred your tree, chop off the branches and lay the boughs over your perennial garden to protect the flowers from the cold.

The Hot and Humid South

~

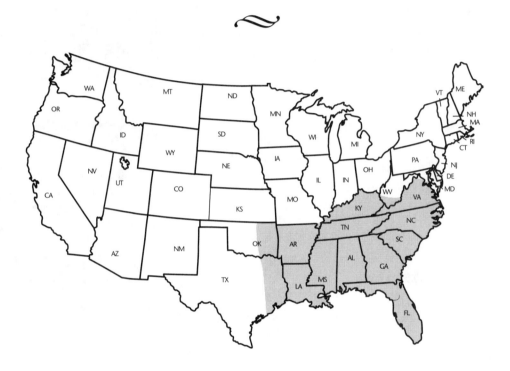

This region begins at the Potomac River and heads west to Little Rock and Oklahoma City and south to New Orleans. It includes south Florida with some variations and also Atlanta, Nashville, Charlotte, and all the states of the Old South.

Here temperatures are generally very hot in summer and a little bit cooler in winter, except in the mountain regions, which are more moderate. Warm-season grasses like Bermuda, St. Augustine, zoysia, and Bahia grow well here, although cool-season grasses like Kentucky blue, perennial rye, and fine fescue also grow quite well along the northern tier of this region and in some of the higher elevations.

The soils here are generally acid and are composed of sand, clay, and loam. The growing season is long, and rainfall is generally adequate.

WHAT TO WATCH FOR

Some areas might actually be getting a little snow this month. Avoid using harsh de-icing chemicals or rock salt on your sidewalks, because both of those could damage your lawn. Try shoveling the snow and then sprinkling a little sand on the walks for traction.

MOWING AND MAINTENANCE

If you have overseeded your summer lawn with winter grass, you are busy mowing. Remember you can mow these grasses at a height of up to 3 inches.

Otherwise, be sure to get your mower blades sharpened if you haven't already, because it won't be long before it's time to cut your summer lawn.

Have your mower blades sharpened at least twice a year. Thanksgiving—at the end of the mowing season—and Fourth of July—the middle of the mowing season—are two good times to get the job done. Dull blades beat your grass to death, leaving it susceptible to disease and drought.

FERTILIZING AND SOIL BUILDING

Do not add any fertilizer this month. If you live in a part of the Southeast where the ground is partially frozen, just sit tight until warm weather arrives. If your ground is warm, here are two things to do:

Take a soil test for your lawn. If you need to add lime, which many lawns in the South lack, now is a good time to do that.

If you don't have a compost pile started already, now is a good time to get one going. Please read the "Clean Easy Compost" chapter in this book for full details.

Top-dress your lawn with organic matter to increase the fertility of the soil and improve the microbial activity. Simply spread ¼ to ½ inch of sifted compost, dehydrated manure, or peat moss, or a combination, over your lawn and rake it in.

Whether your ground is warm or partially frozen, keep adding kitchen scraps and yard debris to your compost pile and turning the contents once a month.

WEED CONTROL

If you have cool-season grass, dig out any Bermuda grass along with dandelions and plantain.

PLANTING AND OVERSEEDING

No planting or overseeding this month. It's still too early. Now is a good time to take a lawn survey to determine if you have any bad spots that need to be repaired when warm weather returns.

WATERING

You won't be watering your lawn this month, but it is a good idea to keep track of what moisture you are getting. Set out a rain gauge or check the precipitation in the newspaper or on the weather programs on radio and TV.

FEBRUARY

WHAT TO WATCH FOR

Once again snow might be a problem in some of the higher elevations, but most areas will see the advent of spring, with crocus and daffodils blooming later in the month.

MOWING AND MAINTENANCE

Many Deep South lawns will begin to green up this month. Bermuda grass can be cut to a height of 2 to 2 ½ inches, while Bahia and St. Augustine grass can be cut as high as 3 inches, so set your mower to cut at this height.

You may be able to rake your lawn and give it an early spring cleanup this month. Raking clears any rocks, fallen leaves, twigs, and other debris that might have accumulated in your yard over the winter and helps aerate the lawn at the same time.

FERTILIZING AND SOIL BUILDING

Don't add any fertilizer at this time.

You can still get a soil test done, and add any lime if necessary. You can also top-dress with compost or other organic matter this month. Keep making compost.

WEED CONTROL

Keep digging out any annual or perennial weeds.

PLANTING AND OVERSEEDING

You can't plant any seeds this month, but you should start shopping for the seeds you want to plant a little later in the spring.

WATERING

No need to water this month because early spring rains should be giving your lawn plenty of water.

WHAT TO WATCH FOR

All those brown Bermuda and zoysia grass lawns will be greening up this month along with the cool-season lawns in the northern areas. It's already summer in the Deep South.

MOWING AND MAINTENANCE

The grass in most of the South should be starting to grow this month. Be sure your mower is in good repair and the blades are sharp.

Cut Bahia, St. Augustine, Kentucky blue, fine fescue, and perennial rye 3 inches tall. Cut Bermuda and zoysia 2 to 2½ inches tall. Cut your grass with a mulching mower or side-discharge mower and leave the clippings on the lawn for a free source of valuable fertilizer and organic matter. Follow the one-third rule and try not to let the grass get too tall before you cut it. If your grass does get too tall, rake up the extra clippings when you cut it and put them in your compost bin.

FERTILIZING AND SOIL BUILDING

It's good to aerate your lawn once the grass gets growing well. Aerate your lawn with a power aerator, a hand-held aerator tool, a forked spade, or spiked golf shoes. Aerating helps water, moisture, and nutrients get deep into the soil to break up compaction, a prime cause of poor grass and the spread of most weeds.

WEED CONTROL

Start digging out weeds by hand to stop their spread. Taller grass will help control most weeds.

PLANTING AND OVERSEEDING

Between now and June is the best time to plant a new lawn with warm-season grasses such as Bermuda, St. Augustine, and zoysia. Be sure to have the soil tested, add any lime if required, add 2 inches of organic matter, and dig the soil to a depth of 4 to 6 inches. Broadcast the seeds or plant the sprigs, plugs, or sod, and keep everything well watered until germination or the other grasses take hold.

Overseeding for winter grass is done in the fall, but overseeding to repair an existing lawn can be done now.

PEST CONTROL

If you have problems with grubs in your lawn, now is a good time to apply grub-busting beneficial nematodes. Wait until there is no longer any chance of frost in your area and then apply the nematodes, carefully following the manufacturer's directions. Nematodes are alive and very delicate—until they get into the soil, where they become deadly to grubs.

WATERING

The only watering you should do now is if you have started planting any seeds.

WHAT TO WATCH FOR

Lawn care will be in full swing for both warm-season grasses and cool-season grasses. Any rough areas in the lawn can be spot-patched this month.

MOWING AND MAINTENANCE

You probably will be mowing every four days as the grass continues to grow rapidly. You may notice that the grass is growing a little bit slower though, without chemical fertilizers, which overstimulate top growth. If your old gas mower is giving you a hard time, consider buying one of the new cordless electric mulching mowers. The new lightweight push reel mowers are perfect for southern lawns, because they cut the grass at just the right height.

FERTILIZING AND SOIL BUILDING

No fertilizer this month, but you could add top dressing if you haven't already done so.

Your compost should now resemble crumbly brown soil instead of leaves and banana peels or whatever you've been putting in it. Stop adding anything new, turn it one more time, and give it a couple of weeks to finish off. Then empty the bin into your vegetable garden and till it in or spread the compost under your shrubs or bushes or around your flower beds for a perfect mulch.

WEED CONTROL

Keep cutting the grass taller than usual and fighting weeds by hand. Tall grass will prevent the growth of crabgrass.

PLANTING AND OVERSEEDING

This is also a good month to plant a new lawn or overseed an existing one with warm season grasses. Simply get the best quality grass seeds, sprigs, or plugs you can afford and follow the instructions in the chapter on lawn renovation earlier in this book.

PEST CONTROL

Many of the new endophytically treated perennial ryegrasses are very resistant to chinch bugs and sod webworms. If you live in a cooler part of the South, where Kentucky bluegrass is widely grown, you may consider switching to ryegrass this fall when it is time to plant.

WATERING

It is still too early to water even though you may be tempted. Early watering trains the grass roots to dwell too close to the soil surface, which will make them lazy, water dependent, and unprepared for drought later in the summer. Withhold the water and force the roots to dig deep in the soil for moisture.

MAY

WHAT TO WATCH FOR

Trees are blooming, birds are singing, grass is growing, and weeds and diseases will be turning up on lawns that are converting from chemical care to environmental care. If you are seeing problems, get a soil test done and aerate your lawn to break up compaction.

MOWING AND MAINTENANCE

Keep mowing. If your grass looks more beat up than cut after you mow, get your blades sharpened.

FERTILIZING AND SOIL BUILDING

May is the first month that it is good to put down an application of natural organic fertilizer for both cool-season grasses and warm-season grasses. In general, it takes 25 pounds of natural organic fertilizer to feed a 2,500-square-foot lawn. Put down half of your fertilizer this month and save the rest for later in the season.

WEED CONTROL

Bermuda, zoysia, and St. Augustine grasses do a very good job of dominating a lawn and keeping weeds out. Perennial rye, fine fescue, and Kentucky blue will also fight off a lot of weeds if you let the grass grow 3 inches tall. Dig out any perennial weeds like dandelion and plantain by hand.

PLANTING AND OVERSEEDING

You can continue planting grass seed, sod, sprigs, or plugs this month. Keep the newly planted grass watered until it gets growing well in the ground, then lighten up.

PEST CONTROL

Brown patch, dollar spot, and other lawn diseases may begin to show up when the weather really gets warm and the humidity starts to rise. They are the result of a weak lawn caused by too much or not enough fertilizer applied at the wrong time, bad watering practices, and other lawn abuse. The diseases are symptoms of a lawn in need of repair, and spraying them with chemicals does not solve the problem. Let your lawn fight off diseases all by itself by following the ten-point program outlined in chapter one.

WATERING

If you can avoid watering your lawn this month, please do. It is better to hold off and condition the lawn to dig its roots down deep in the soil to search for water.

If you do water, water once a week, spray the water on the lawn as slowly as possible, and soak the ground to a depth of 6 inches.

JUNE

WHAT TO WATCH FOR

The heat is building, but that is fine for Bermuda and the other warm-season grasses. Cool-season grasses may begin to go dormant, which is also fine, because they will return in the fall.

MOWING AND MAINTENANCE

Keep mowing and mulching those grass clippings. Unless you've had a lot of rain, your mowing frequency should be slowing down.

FERTILIZING AND SOIL BUILDING

It's a little late for any fertilizing. If you haven't done so already, you might want to apply half of your natural organic fertilizer to lawns growing Kentucky blue, fine fescue, and perennial rye. Save the main fertilizing for the fall.

WEED CONTROL

Keep up your hand weeding, because dandelions and other spring flowering weeds are at their weakest.

PLANTING AND OVERSEEDING

This is the last month this year for you to successfully plant new grass seeds, sprigs, plugs, or sod. By now, any grass that you planted earlier should be up and growing quite well. Patch any spots that didn't do well.

PEST CONTROL

Mole crickets (found only in the South), sod webworms, and chinch bugs may start doing damage to overfertilized, compacted lawns with thatch buildup. You are already fighting those pests and most diseases by aerating your lawn, feeding it with natural organic fertilizers, and watering carefully. You might also consider replanting your lawn with endophytically treated varieties of perennial rye or tall fescue, which repel those three bugs.

If you are having problems with chinch bugs, which are $1/4$-inch-long black bugs with white wings that feed on the tops of your grass, primarily St. Augustine grass, switch to the Floratam variety of St. Augustine grass next year.

WATERING

Don't water any more than once a week, and less if you've been getting good rainfall. Use an impulse sprinkler or an improved oscillating sprinkler to get the most water to the grass. Water early in the day or late in the afternoon so you won't lose any moisture to evaporation.

An evenly applied weekly watering program will help your lawn fight off many bugs and diseases that attack drought-stressed lawns.

WHAT TO WATCH FOR

The big push of springtime activities is starting to subside. It's a good time to take a long look at your lawn. Is the grass growing well? Do you have bare spots? Is there too much shade or not enough in your yard? Make an assessment of any problems you might have and any changes you might want to make.

MOWING AND MAINTENANCE

If you've been mowing a lot this year, it won't hurt you to get your blades sharpened. Dull blades chew up the grass rather than cut it, which can damage the grass and promote weed growth.

FERTILIZING AND SOIL BUILDING

No fertilizing or soil building this month, except that you should continue to make compost with excess grass clippings, spent plants, kitchen scraps, and other good things. This warm weather will really speed up the process of composting.

WEED CONTROL

Keep your lawn mower blades sharp and cut the grass high to force out most weeds. Sometimes weeds or grasses like to spread to areas like your flower beds, where you don't want them to go. Pull them out by hand or spray them with a nontoxic spray often sold under the name Sharpshooter.

PLANTING AND OVERSEEDING

Don't plant any grass this month. It's too hot, and the weed seeds in your lawn are beginning to germinate.

PEST CONTROL

Chinch bugs, mole crickets, and other grass-eating bugs may be appearing on your lawn. Most bugs are attracted to lawns that are heavily compacted and overfertilized and have a good amount of thatch buildup. The long-term solution of reduced use of chemicals, restricted water use, and top dressing to eliminate thatch will work the best. In the short term, try spraying affected areas with insecticidal soap, which you can find at your lawn-and-garden center.

WATERING

It is normal for the cool-season grasses, Kentucky blue, perennial rye, and fine fescue, to go dormant and turn slightly brown during July and August. Don't water them. As soon as the rain returns in September, the grass will turn green and begin to grow. If you want green grass all summer long, and you can afford the water, water no more than once a week and water the ground thoroughly to a depth of 4 to 6 inches. Remember, you're not watering the grass leaves, you're watering the roots.

WHAT TO WATCH FOR

Cool-season grasses like Kentucky blue have probably gone dormant on your lawn by now. That's okay. They will return to green as soon as the rains return and the temperature cools off.

MOWING AND MAINTENANCE

Keep mowing every four days or so for warm-season grasses. Unless you're watering heavily, cool-season lawns will require very little mowing this month.

FERTILIZING AND SOIL BUILDING

Now is a good time to apply the second half of your natural organic fertilizer left over from spring for your warm-season lawn. If you haven't added any yet, just apply half of your 25-pound bag now and save the rest for spring.

WEED CONTROL

Dig out weeds by hand and keep the grass growing at its optimum height to force out weeds.

PLANTING AND OVERSEEDING

Late August is a good time to overseed a warm-season lawn with perennial ryegrass or Kentucky bluegrass for a green winter lawn. Be sure to ask your lawn-and-garden center for the best new disease- and pest-resistant varieties. Simply cut your lawn to 1 inch tall or lower. Rake the lawn vigorously with a steel garden rake to rough up and expose the soil. Broadcast the seeds and then lightly rake the lawn again to be sure the seeds get in good contact with the soil. Sprinkle a ¼-inch layer of compost, topsoil, or sand, or a combination, over the area and water. Keep the lawn lightly watered until germination. As your warm-season grass goes dormant for the winter, your lovely winter grass will start to grow and take over.

PEST CONTROL

Grubs will be coming back into the lawns this month. Now is a very good time to apply grub-busting beneficial nematodes. They will live and continue to kill grubs until frost, if you get frost in your area.

WATERING

Continue your sparse watering program.

SEPTEMBER

WHAT TO WATCH FOR

Cool evenings at least, and as a little rain begins to return to the upper South, your cool-season grass is beginning to grow again. Warm-season grasses are beginning to slow down.

MOWING AND MAINTENANCE

Keep mowing and recycling those grass clippings. If you have a bagging attachment for your mower, make sure you can find it in the garage because you are going to need it soon.

FERTILIZING AND SOIL BUILDING

If you haven't emptied your compost bin, be sure to do it this month. You can spread it around your perennial beds as mulch or dig it in to parts of your vegetable garden to improve the soil. You can even put it in the pots of your indoor plants to make them healthier, too.

If you don't have a bin, buy one now. Look for a sturdy, hard-plastic, wood, or plastic-coated metal bin that measures approximately 3 feet by 3 feet by 3 feet. Place it on a level surface in an inconspicuous place that gets full sun or partial shade.

WEED CONTROL

Keep digging those weeds out by hand.

PLANTING AND OVERSEEDING

This is another good month to overseed your warm-season lawn with cool-season grass.

If you live in one of the areas of the South where you can grow cool-season grasses, like Kentucky blue, perennial rye, and fine fescue, in the summer, now is the very best time to plant grass.

For a completely new lawn, simply take a soil test and add lime or sulfur if necessary, apply 2 inches of organic matter and a good high-phosphorus natural organic starter fertilizer, and dig the whole lawn to a depth of 4 to 6 inches. Smooth out the lawn with a rake or a roller, broadcast the grass seed, rake or roll the lawn again, cover with straw, and keep watered until germination in one to two weeks.

This is also a good time to spot-repair or overseed a cool-season lawn with more seed.

PEST CONTROL

Pests should not be much of a problem in most parts of this zone, as the weather begins to cool off this time of year. Grubs are still hatching and may well begin to chew on your lawn. A recent university study found that walking on the lawn with spiked shoes when grubs are near the surface kills almost as many grubs as any pesticides, natural or chemical, do. You might give it a try.

WATERING

If you've been watering once a week, keep up this program, but reduce the amount if you start getting rain.

OCTOBER

WHAT TO WATCH FOR

October is a very pretty month in the South as the heat of summer starts to subside. Leaves of hardwood trees like maple and oak will begin to fall this month.

MOWING AND MAINTENANCE

As the first leaves begin to fall, mow them into your lawn for a free source of fertilizer. Most mowers, mulching mowers being the best, will finely shred a light dusting of leaves and return them to the lawn, where they become a valuable source of phosphorus fertilizer and organic matter.

FERTILIZING AND SOIL BUILDING

Now is a good time to apply fertilizer for cool-season grasses like Kentucky blue, perennial rye, or fine fescue, whether you are overseeding them as winter grass or growing them in the summer.

WEED CONTROL

Keep pulling weeds out by hand.

PLANTING AND OVERSEEDING

If you are going to overseed your Bermuda grass with Kentucky blue or perennial rye, now is the time to do it. Cut the Bermuda grass down to 1 inch. Rake the lawn to get up any debris and help scratch the soil surface. Sprinkle the seeds evenly over the lawn, and water it lightly every day until the new grass germinates and starts to get green.

WATERING

Water only to help germination of newly seeded lawns.

NOVEMBER

WHAT TO WATCH FOR

The big leaf fall and cooler weather.

MOWING AND MAINTENANCE

This is probably the last month that you will be cutting grass. Be sure to lower your mower's cutting height by 1 inch for the last mowing, so that your tall grass does not mat over during the winter, which might cause diseases to spread next year.

FERTILIZING AND SOIL BUILDING

You can still add natural organic fertilizer if you haven't already done so. When you do get leaf fall, mow some of the leaves right into the lawn, where they become a valuable source of phosphorus fertilizer. Rake or shred the leaves, or, better yet, mow them into your mower's bagging attachment and place them in your compost bin. By spring, they should be compost.

This is also a good month to add any lime if your lawn needs it and to top-dress with a ¼-inch layer of organic matter.

WEED CONTROL

Dig out any weeds by hand.

PLANTING AND OVERSEEDING

You can still overseed for a winter lawn in most parts of the upper South.

DECEMBER

WHAT TO WATCH FOR

Unless you've overseeded, there isn't much going on in the lawn. Try not to tear it up too much with your outdoor Christmas decorations.

MOWING AND MAINTENANCE

If you overseeded your lawn for a green winter, you will keep mowing this month. Otherwise, make sure you have changed the oil and run the gas out of the tank for winter storage.

FERTILIZING AND SOIL BUILDING

No fertilizer this month. Keep making compost with kitchen scraps and other yard debris. Turn the pile once a month. Compost will continue to happen as long as air temperature is above 40 degrees.

WATERING

No need to water, but you should be sure that all the water is out of your hoses before you hang them up in the garage for the winter.

The Hot and Dry
Southwest

~

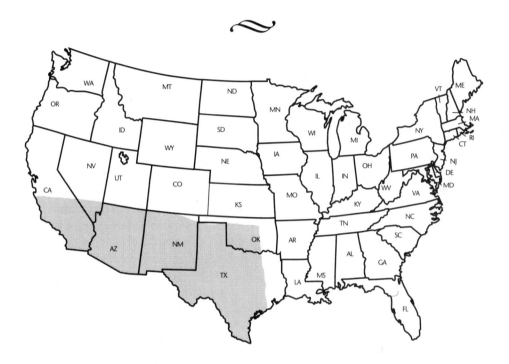

This is a warm and dry part of the country that begins somewhere around Dallas and Fort Worth in Texas and stretches west to Los Angeles. It includes West Texas from San Antonio to Abilene, Lubbock, Odessa, and El Paso, and Albuquerque, Las Cruces, Phoenix, Las Vegas, and Southern California, including the valley towns as far north as Sacramento.

The soil is either sand or clay, usually neutral pH or slightly alkaline. It's a good idea to get a soil test done for your lawn. Warm-season grasses like Bermuda, St. Augustine, and zoysia are the most commonly grown. The growing season is long, but rainfall can be a problem, with some areas getting as little as 10 inches per year.

There are lots of local climatic variations, so please be sure to ask your local County Extension agent or lawn-and-garden center for lawn advice suitable for your area.

Please read chapter one, "Top Ten Ways to a Beautiful Easy Lawn," to ground you in my environmental lawn care program before you follow the steps in this chapter.

JANUARY

WHAT TO WATCH FOR

It's not likely that you will be getting snow this month, although it does happen occasionally in this region. If you do get some, avoid using rock salt or chemical de-icers to clear it. Instead shovel the area and sprinkle some granulated natural organic fertilizer on the walk. The fertilizer will actually benefit the grass in the spring.

MOWING AND MAINTENANCE

Since it is a little early to begin mowing, take your mower blades in for a good sharpening if you didn't get to it last fall. You might also get your mower tuned up, so it is ready when the grass does begin to grow.

This is a good month to shop for a new mower. Be sure to look at the cordless electric mowers, the lightweight reel mowers, and the mulching mowers with a bagging attachment.

FERTILIZING AND SOIL BUILDING

You lucky people can make compost all year round. Please read the "Clean, Easy Compost" chapter in this book and get a bin going in your backyard. Be sure to add kitchen scraps, paper plates, paper towels, and paper napkins to the bin along with any yard debris that accumulates.

People in the extreme southern zones of this region can prepare to start adding fertilizer or planting grass seed, although fall is really the better time to do both of these.

Take a soil test. Many soils in this region are alkaline and may need the addition of gypsum to neutralize the soil.

You might also take the time now to top-dress your lawn with organic matter, increasing the fertility of the soil and improving the microbial activity to prevent diseases and make the lawn more drought resistant. Simply spread over your lawn a $1/4$- to $1/2$-inch dusting of sifted compost, dehydrated manure, or peat moss, or a combination, over your lawn and rake it in. This top dressing alone may correct any alkaline soil imbalance you have.

WEED CONTROL

Start digging out weeds by hand as they appear.

PLANTING AND OVERSEEDING

It's a little early to start planting grass seed except in the most southern areas. Instead, take a lawn survey to search out and identify any bad spots that need to be repaired as soon as weather permits.

WATERING

You won't be watering your lawn this month, but it might be a good idea to keep track of what precipitation you are getting. Get your information from the newspaper or TV weather person, or install a rain gauge of your own.

FEBRUARY

WHAT TO WATCH FOR

Grass will start growing later this month, and many beautiful flowers will come into bloom.

MOWING AND MAINTENANCE

Get your lawn mower blades sharpened if you haven't already. Dull blades demolish your grass rather than cut it, which can lead to disease and weakened grass.

Now is the time to give your lawn a spring cleaning, which aerates the soil by clearing out any rocks, fallen leaves, twigs, or other debris that might have accumulated over the winter.

Set your mowing height according to the type of grass you are growing. Cut tall fescue, buffalo, and St. Augustine grass at 3 inches tall, Bermuda and zoysia at 2 inches. No one should be growing Kentucky bluegrass in this region, because it demands too much water and fertilizer for the hot, dry weather.

FERTILIZING AND SOIL BUILDING

It is still a little early to add any fertilizer for most areas. Have your soil tested, top-dress with organic matter, and keep making compost.

WEED CONTROL

Keep digging out those weeds as they emerge.

PLANTING AND OVERSEEDING

Now is a good time to prepare to plant seeds for this season. Go to your local lawn-and-garden center and make sure they have the types of grass seed you want. Be sure to buy premium-quality grass seeds and other products to get the best results.

WATERING

Hopefully you will be getting enough early spring rains to replenish your soil's moisture levels.

WHAT TO WATCH FOR

March is the transition month, when the warm-season grasses start to green up. Get ready to mow.

MOWING AND MAINTENANCE

Have your mower in good repair and start mowing. Remember to cut buffalo, tall fescue, and St. Augustine grass at the 3-inch level and zoysia and common Bermuda at 2 to $2\frac{1}{2}$ inches tall. Tall grass helps develop a stronger grass plant that can fight off many pests, diseases, and drought all by itself. Cut your grass with a mulching or side-discharge mower and leave the clippings on the lawn for a free source of valuable fertilizer.

Follow the one-third rule and try not to let the grass get too tall before you cut it. If you get behind and your grass does get too tall, be sure to rake up the clippings and put them in the compost bin.

FERTILIZING AND SOIL BUILDING

If your compost pile froze over an unusually harsh winter, it should be thawed by now. Turn it and water it if it has dried out.

WEED CONTROL

You are already controlling crabgrass by letting your grass grow taller. Tall grass creates shade, which prohibits the germination of sun-loving crabgrass seeds. University studies back this up, so give it a try.

PLANTING AND OVERSEEDING

Between now and June is a favorable time to plant a new lawn with warm-season grasses, including Bermuda, St. Augustine, and zoysia. Please read chapter three on lawn renovation for complete details on grass types and grass planting. Be sure to have the soil tested, add 2 inches of organic matter, and till the soil to a depth of 4 to 6 inches. Broadcast the seed or plant the sprigs, plugs, or sod, and keep everything watered until germination or the new grasses take hold.

WATERING

Water only the grass seeds, sprigs, plugs, or sod you have planted.

WHAT TO WATCH FOR

All of the grasses are up and growing this month, and spring is fully under way.

MOWING AND MAINTENANCE

Your grass will be growing a little bit slower this year if it's the first year you didn't use high-nitrogen, quick-release synthetic fertilizers. Natural organic fertilizers grow the grass slower, meaning you don't have to cut it as often.

FERTILIZING AND SOIL BUILDING

Your compost should now resemble crumbly brown soil rather than banana peels, leaves, or other debris. Empty the bin and spread the matter over your lawn.

WEED CONTROL

Your taller grass should be holding down the crabgrass. You can continue to dig out dandelion and other perennial weeds by hand.

PLANTING AND OVERSEEDING

You can continue planting, overseeding, or spot-patching your lawn with improved grass seed varieties. Keep the young starts watered until they get established.

PEST CONTROL

If your lawn is showing any signs of diseases, you should probably plan to replant with disease-resistant strains of your favorite grass type.

WATERING

Water new grass only.

WHAT TO WATCH FOR

Now that the heat of late spring is fully upon you, the grass is growing and weeds and diseases might be turning up on lawns that are converting from chemical care to environmental care.

MOWING AND MAINTENANCE

Bermuda, St. Augustine, and buffalo grasses are tougher to cut than their northern brethren. Get your blades sharpened again if your grass looks more beat up than cut after you mow.

FERTILIZING AND SOIL BUILDING

This is the month when you should fertilize your lawn. In general, it takes 25 pounds of natural organic fertilizer to feed a 2,500-square-foot lawn. Spread half of your fertilizer now and save the rest for later in the season.

Compaction is a big problem on lawns that get a lot of use. People walking or running on the lawn will pack down the soil, making it more difficult for plants to grow there. You can overcome compaction by aerating your lawn with a power aerator or a hand-held aerator tool, or by walking on the lawn with spiked golf shoes. Give it a try. Aeration also helps water and fertilizer seep deeper into the lawn, where it can really do some good.

WEED CONTROL

The type of grass you are growing, Bermuda, St. Augustine, or zoysia, forms a deep thick turf, which does a good job of keeping weeds out of the lawn. Rather than spend time and money spraying your lawn with herbicides, you are better off improving the fertility and water-retentive qualities of your lawn.

PLANTING AND OVERSEEDING

Continue planting grass seed, sod, sprigs, or plugs this month, and keep the newly planted grass watered until it is growing well. Then cut back on your watering.

PEST CONTROL

Because most lawn diseases are associated with humid weather, you are not as prone to getting them as the southern states farther east. The best way to control diseases is to prevent them; chemicals only solve disease problems for a short time. You can fight off diseases by building your lawn's soil with natural organic fertilizer, top dressing with organic matter, and planting disease-resistant grass varieties.

WATERING

Watering lawns in the dry Southwest is always a question of priorities. Grassy lawns are not necessarily a natural part of your landscape. If you do water, make every little drop count. Water slowly with a misty spray from an in-ground watering system or drip hose turned on its back, which will also emit a spray. Water no more often than once a week and then for about two hours, so the water can seep down deep into the soil. If you get any water runoff, you are watering too fast, turn the spigot to a lower setting.

JUNE

WHAT TO WATCH FOR

The heat is building, but your Bermuda, buffalo, St. Augustine, or zoysia grass lawn prefers it that way.

MOWING AND MAINTENANCE

Unless you've had a lot of rain, your grass should be growing a little slower by now. Keep mowing and mulching those grass clippings.

FERTILIZING AND SOIL BUILDING

It is not too late for fertilizing if you use natural organic fertilizer. Naturals break down more slowly and won't burn the lawn.

It's never too late to make compost. Add any excess grass clippings to the bin along with kitchen scraps and turn once a month. Add a bucket of water now and then if the pile seems to be drying out.

WEED CONTROL

Keep pulling weeds by hand.

PLANTING AND OVERSEEDING

This is the last month this year for you to successfully plant new grass seeds, sprigs, plugs, or sod. By now any grass that you planted earlier should be up and doing well.

PEST CONTROL

Good lawn maintenance is the best way to fight pests in the long run. Environmental lawns like you are growing can fight off most pest infestations in this part of the country.

WATERING

Another way to cut down on the amount of water your lawn requires is to cut down on the size of your lawn. Consider planting some ground flowers or shrubs that are native to your area in parts of your yard instead of growing grass. Native plants in your area must be able to withstand periods of very little water. Contact your local nursery or Cooperative Extension office for advice.

WHAT TO WATCH FOR

It is a good time to make an assessment of your lawn, now that the big push of springtime activities has subsided a little bit. Is the grass growing well? Are there bare spots? Wouldn't it be nice to plant a tree and create a little more shade from that hot summer sun?

MOWING AND MAINTENANCE

Get your mower blades sharpened again. Most Cooperative Extension offices recommend blade sharpening twice a year now, because dull blades are murder on your grass.

FERTILIZING AND SOIL BUILDING

If you mow with a mulching mower or simply leave the clippings on the lawn, you are always building the soil. The clippings disappear in your tall grass and decompose in a matter of days, providing you with half of your fertilizer needs for free.

WEED CONTROL

Keep your mower blades set at 3 inches for buffalo and St. Augustine grass and 2 inches for Bermuda and zoysia to prevent weeds from penetrating your turf.

PLANTING AND OVERSEEDING

Don't plant any grass this month except in emergencies caused by landscaping changes.

PEST CONTROL

Most bugs are attracted to lawns that are heavily compacted and overly fertilized and have a lot of thatch buildup. Reducing the use of chemicals, restricting water use, and top dressing to promote microbial activity and eliminate thatch will work the best over the long term. In the short term, try spraying affected areas with insecticidal soap, which you will find at your lawn-and-garden center.

WATERING

To help relieve your need to water, most lawn care professionals will tell you to let your grass grow a little taller during the very hot summer months. Taller grass seems to withstand the dry heat better than short grass. If you haven't done so already, raise your mower blades to 3 inches tall for St. Augustine or guffalo grass and 2 to 2 1/2 inches for Bermuda and zoysia.

AUGUST

WHAT TO WATCH FOR

Watering restrictions may put a crimp in your lawn plans this month, but a healthy lawn will survive dry conditions.

MOWING AND MAINTENANCE

Keep up with your mowing program and remember to follow the one-third rule. Never cut off more than one-third of your grass's height at any one time. Deeper cutting at this time could cause scalping, a condition that can damage your lawn's roots.

FERTILIZING AND SOIL BUILDING

Hold off on any additional fertilizing until the weather cools a little bit next month.

Compost making should be at its prime this month with warm temperatures and plenty of kitchen scraps and garden debris to add to the pile.

WEED CONTROL

Dig out any weeds by hand.

PLANTING AND OVERSEEDING

Wait until slightly cooler temperatures next month to plant grass.

PEST CONTROL

If you find you are having grub damage, now is a good time to apply beneficial nematodes to control them. If your lawn is showing signs of spotty damage, and you know that grubs are a problem in your area, check for grubs like this: Dig out a 1-square-foot patch of turf. If you find more than eight or ten grubs living there on top of the soil, you may have a grub problem.

Apply beneficial nematodes according to the manufacturer's instructions.

WATERING

Frequent light watering is the worst thing you can do for your lawn. Instead, water deeply to soak the ground to a depth of six inches and do it only once a week. Shallow watering promotes shallow root growth, which makes your lawn even more dependent on shallow watering.

WHAT TO WATCH FOR

This is the best month for planting new grass. Get ready.

MOWING AND MAINTENANCE

Continue mowing and recycling those clippings by leaving them on the lawn or at least putting them in your compost bin.

FERTILIZING AND SOIL BUILDING

Now is a good time to apply a full dose of natural organic fertilizer or the remaining half from last spring. Use a spreader if you have a large lawn, or broadcast the material by hand over a small one.

This is also a good month to aerate your lawn and top-dress with natural organic matter.

WEED CONTROL

Weeds such as crabgrass like to sprout this month. Keep the grass cut higher to prevent weeds from getting a foothold.

PLANTING AND OVERSEEDING

This is a good month to plant a whole new lawn or overseed an existing one. For a new lawn: Till the old one and remove any lawn debris. Add an inch or two of organic matter, a dose of high-phosphorus fertilizer, and till again. Rake smooth and plant the grass seed, sprigs, sod, or plugs. Cover the seeds with straw or top soil, to keep the soil damp, and water lightly every day until the grass is up and growing. To overseed an existing lawn, mow the grass as short as possible, rake the ground to scar it, spread on the same organic matter and fertilizer, sprinkle on the grass seed, cover with topsoil or straw, to keep the ground moist, and water often until the grass sprouts. Please read chapter three on lawn renovation for more details.

WATERING

Water only if you have newly planted grass.

OCTOBER

WHAT TO WATCH FOR

As the blazing heat of the summer subsides, the Southwest becomes even more beautiful. Leaves of hardwood trees will begin to fall this month. Time to make compost.

MOWING AND MAINTENANCE

You may be growing trees that lose their leaves in the fall. It's nothing like Vermont, but you may get some. Mow some of those leaves with your mulching mower and return them to the lawn, where they become a valuable source of phosphorus fertilizer and organic matter.

FERTILIZING AND SOIL BUILDING

You can continue fertilizing with natural organic fertilizer this month. The roots of the plants will be storing the nourishment for the long winter ahead.

Make more compost. Gather up any leaves, spent flowers, and garden residue, and put them in the bin.

WEED CONTROL

Keep pulling weeds out by hand before they get a chance to take over.

PLANTING AND OVERSEEDING

Hold off on planting and overseeding until next year.

WATERING

Hopefully the fall rains will have started, and you won't need to water.

NOVEMBER

WHAT TO WATCH FOR

Cooler weather means your lawn will start to slow down.

MOWING AND MAINTENANCE

As the mowing season slows down, be sure to drain the oil out of your lawn mower and let the gas run out, too. Even though you don't get really cold winters, leftover gas and oil can clog up your mower and cause problems next year.

Cut your grass a little bit lower for the last mowing to prevent matting over the winter. Set the height at 2 inches for buffalo and St. Augustine grass and 1 inch for Bermuda and zoysia.

FERTILIZING AND SOIL BUILDING

Hold off on all of this until next spring.

WEED CONTROL

Keep pulling weeds by hand.

DECEMBER

WHAT TO WATCH FOR

Snow birds and Santa Claus.

MOWING AND MAINTENANCE

Be sure your mower is cleaned and ready for next year. Make sure your cordless electric mower is plugged in to keep the batteries fresh for next year.

FERTILIZING AND SOIL BUILDING

No fertilizing or soil building this month.

Keep making compost. Shred your Christmas tree and add it to the compost bin. Rake and shred any other yard waste and add it to the bin, too. Compost will continue to be produced as long as the air temperature is above 40 degrees. Turn the pile once a month.

WATERING

No need to water this month. You might think about asking for a new hose for Christmas if you need one for next year.

The High and Dry Great Plains

❧

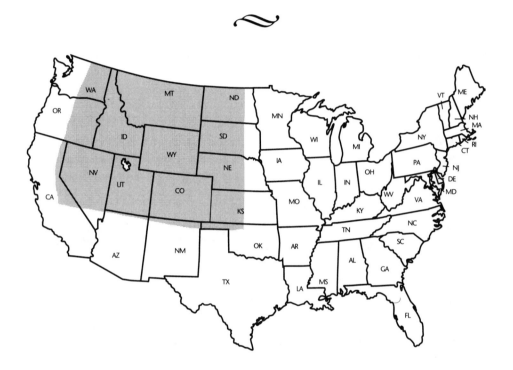

This is an area that begins in the Texas Panhandle, heads west to Santa Fe, north through Reno and Spokane, and east to Denver and the Dakotas. It includes Salt Lake City, Casper, Great Falls, and Boise.

The area here varies from the hot and dry Great Plains in the eastern part, where native grasses such as buffalo grass are good to grow, to the slightly cooler mountains in the west, where cool-season grasses such as Kentucky bluegrass, fescue, and perennial rye can be grown. The soil is generally alkaline or neutral, and rainfall is usually in short supply.

Please read and refer to chapter one, "Top Ten Ways to a Beautiful Easy Lawn," to ground you in my environmental lawn care program before you follow the steps in this chapter.

JANUARY

WHAT TO WATCH FOR

Cold winter weather will have set in, and most areas will have a good blanket of snow, especially in the higher elevations. Snow is good for your lawn because it acts as a blanket to protect the turf from freezing winds.

Salt and other de-icing chemicals can damage your lawn. Try clearing your own walkways with sawdust, kitty litter, or sand instead of rock salt. You could even be helping your lawn if you used a dry granular natural organic fertilizer as a de-icer.

MOWING AND MAINTENANCE

You won't be mowing any grass this month, but you could prepare for the upcoming season by getting your blades sharpened. Make sure your continuous-charge electric lawn mower is plugged in, so it will be ready when the weather changes.

FERTILIZING AND SOIL BUILDING

Shred your Christmas tree and add it to the compost bin along with kitchen scraps and household paper products. Your pile will be frozen this month, but you can still add things to it. If you don't already have a compost pile, this is as good a month as any to get one started. Please read the "Clean, Easy Compost" chapter in this book.

Now is a good time to take a window-shopping trip to your lawn-and-garden center to make sure they plan to stock your choice in natural organic fertilizer, grass seed, and soil test kits.

WEED CONTROL

Get outside and dig dandelions and plantains if you live in a southern county and are having a mild winter.

PLANTING AND OVERSEEDING

Now is a good time to start thinking about your grass. If you have been having problems with diseases and drought stress, you may want to replant your lawn this spring or next fall. Be sure your lawn-and-garden center is going to carry the new improved blends of endophytic perennial rye, fine fescue, and disease-resistant Kentucky bluegrass. Better yet, ask your garden center if the new varieties of buffalo grass are going to be available.

WATERING

Rain and snow should be providing all the moisture you need. But amounts can vary. Get a simple rain gauge and keep track of how much water you are getting right in your own backyard.

WHAT TO WATCH FOR

It is not likely that you are going to see any grass growing this month. But you can keep making compost.

MOWING AND MAINTENANCE

Now is a good time to shop for a new lawn mower, when sales are slow and dealers are ready to deal. Consider one of the mulching mowers, a cordless electric mower, or maybe one of the lightweight reel mowers.

Take your mower in for a tune-up while the repair shops are less busy. A gas-powered mower needs a tune-up every year.

FERTILIZING AND SOIL BUILDING

If you get any days above 40 degrees, your compost pile may thaw out enough for you to turn it over. Continue to add kitchen scraps and any yard debris you might find.

PLANTING AND OVERSEEDING

If you plan to reseed your lawn this year, you might as well buy the seeds you want while they are in stock, so you'll have them at the ready.

PEST CONTROL

If you have a problem with grubs, you might want to make sure your lawn-care center will have a supply of grub-busting beneficial nematodes on hand later in the spring.

WATERING

No need to water this month, but do keep track of your precipitation in your rain gauge.

WHAT TO WATCH FOR

March weather is always very unusual. You could be buried in snow one day and have spring-like weather, rain, and mud the next. This means that the lawn season is just about to start.

MOWING AND MAINTENANCE

Before you start mowing, give your lawn a good raking, which I call a spring lawn cleanup, to gather any dead limbs, leaves, and other debris that can damage the grass and give weeds an opportunity to sprout.

Be sure your mower's blades are sharpened and that your mower is tuned up or charged up and ready to go.

Set the mower's cutting height to 3 inches and start mowing once the grass starts to grow.

FERTILIZING AND SOIL BUILDING

Mid- to late March is a good time to take a soil test. Simply take your soil samples to your Cooperative Extension office or do the analysis yourself. If you need to add some gypsum to neutralize your alkaline soil or lime to neutralize your acid soil, now is a good time to do it.

Fall is the best time to fertilize your lawn, but spring is second best. Buy your fertilizer now, but don't put it on until after your grass has gotten its first good green growth spurt.

Top-dress your lawn by sprinkling on and raking in ¼ to ½ inch of sifted compost, dehydrated manure, or peat moss. Top-dressing is a good way to fight thatch and other lawn problems. Now is a good time to do it, once the ground has thawed.

WEED CONTROL

Pull out knotweed, chickweed, and ground ivy and dig out perennial weeds such as dandelion and plantain as they emerge. Cutting your grass at a 3-inch height is the best way to prevent the emergence of annual weeds such as crabgrass.

PLANTING AND OVERSEEDING

You can begin planting cool-season grasses in the southern parts of this region as soon as air temperatures are consistently in the 70- to 75-degree range. Always plant a mixture of endophytic perennial ryegrass, fine fescue, and disease-resistant Kentucky bluegrass that is blended for either sun or shade, depending on your conditions. Consider planting the new improved varieties of buffalo grass in a sunny open yard. Buffalo grass does not tolerate shade very well.

PEST CONTROL

Pests should not be causing any problems this month, but grubs, if you have any, will begin migrating to the surface as soon as the weather warms.

WATERING

You should be getting so much rain right now that you do not have to water your lawn.

WHAT TO WATCH FOR

Most areas will have their last frosts this month, and spring will be under way. The daffodils and crocus will be blooming, and the grass is growing.

MOWING AND MAINTENANCE

Your grass will be growing at a good rate now, and you are probably mowing as often as every four or five days. Remember to follow the one-third rule and never cut off more than one-third of your grass's height at any one time. If you fall behind and the grass is too tall to cut with a mulching mower, you may want to attach the bag and collect the clippings and then put them in the compost bin. If you have a side-discharge mower and you feel the grass clippings are too dense, rake them and put them in the compost bin.

FERTILIZING AND SOIL BUILDING

If you are going to add natural organic fertilizer to your lawn this season, now is the time to do it. In general, a 2,500-square-foot lawn will need about 25 pounds of natural organic fertilizer each year. Apply half now and add the rest in the fall.

WEED CONTROL

Let your grass grow 3 inches tall, and you will be able to prevent the growth of crabgrass and most other annual weeds. Tall grass shades the ground where crabgrass seeds need sunlight to germinate.

As you dig out dandelions and other weeds by hand, be sure to spot-seed with new grass seed to let the grass take over where the weeds used to grow.

PLANTING AND OVERSEEDING

Fall is still the better time to plant grass seed because there is no competition with weed seeds, but now is also a good time. Be sure to prepare the soil, add natural organic fertilizer, sprinkle on plenty of grass seed, and keep it watered until it germinates.

PEST CONTROL

After all danger of frost has passed, apply beneficial nematodes according to the manufacturer's instructions. Poor application is the principal reason for failure.

WATERING

It is still too early to start watering. If you start watering now, your grass will develop shallow roots that will make it drought prone later in the season. Watering now could also contribute to the spread of lawn diseases.

WHAT TO WATCH FOR

Your grass should be looking great this month. If not, you may have some problems. If it looks weak and weeds are coming on strong, you need a soil test and may need to add more fertilizer. If things look really bad, plan to replant your lawn with better grass seeds later this summer or fall.

MOWING AND MAINTENANCE

Dull blades beat, tear, and shred your grass, damaging your lawn. You may want to consider sharpening your blades again if this is happening.

FERTILIZING AND SOIL BUILDING

Now that the weather is warm, you can finish off that compost you've been making. Stop adding new materials, turn the pile once a week for one month, and you should have finished compost. Empty the bin and spread it around your flowers as mulch or work it into your vegetable garden. Start a new batch of compost.

Finish up your fertilizer applications now. You are always adding free fertilizer to your lawn as you mulch your grass clippings and let them decompose in the lawn.

A bin full of finished compost gives you free organic mulch—about $50 worth if you had to buy it at a lawn-and-garden center. Buy a sturdy wooden, hard plastic, or metal wire compost bin (a loose pile can blow all over your yard). Place the bin at a level sunny or semi-shady spot, fill it with yard waste and kitchen scraps, and be patient. Compost will take up to six months to happen. It is ready to be spread under your bushes or around your flowers as mulch or raked into your lawn as added soil enrichment when it looks more like soil than banana peels.

WEED CONTROL

Dandelions have just used up all their food reserves to create their flowers, and the plants are at their weakest point. Now is the time to dig them up.

PLANTING AND OVERSEEDING

Hold off on planting any more grass seeds because of the emergence of annual weeds such as crabgrass. Wait until late August to start planting again.

PEST CONTROL

If you have a problem with chinch bugs, now is the time that they will appear. Chemicals will control the problem for only a little while. The best solution is to plan to replant your lawn this fall with perennial ryegrass and fine fescue that have been treated with endophytes that repel chinch bugs.

WATERING

Watering is advised now only in the southern zones of this region and then only if you are getting drought conditions. If you really want to water, water no more than once a week and then deliver about 1 inch of water so that it seeps down to a depth of 6 to 8 inches. This watering can take up to 3 hours, depending on your soil type.

JUNE

WHAT TO WATCH FOR

Now is a good time to take a full assessment of your lawn and yard environment while the trees are fully leafed out and the bushes are or have been in bloom. Your trees may have grown a lot since last time you thought about it, and you may now be trying to grow sun-loving grass in a shady lawn. If your grass is weedy and not growing well, you may want to plan for a renovation this fall. You may want to shrink the size of your lawn by planting some ground covers under trees and along the garage. Now is a good time to plant ground covers such as wild ginger, myrtle, and ivy.

MOWING AND MAINTENANCE

You should still be mowing every four days or so to keep up with your lawn care program as outlined in the first chapter of this book. It will really pay off for you, because you won't be tempted to use chemicals.

FERTILIZING AND SOIL BUILDING

You can still add that layer of top dressing now if you didn't get to it earlier in the season. Don't add any fertilizer at this time.

WEED CONTROL

Now is a good time to assess the amount of weed infestation you have in your lawn. If you have been following the ten-point program featured in chapter one, and you still have serious weed problems, you will want to consider replanting your entire lawn later this season after you've taken a soil test.

PLANTING AND OVERSEEDING

Wait until August to any major replanting or overseeding. The only planting you should do now is spot patching and emergency planting caused by digging or major landscape changes.

PEST CONTROL

Now is a good time to check if you really have a grub infestation, because the grubs are getting ready to emerge as beetles and start flying around for the summer. Dig up a 1-foot-square patch of turf. Count the grubs. If there are more than eight or ten, then you have the potential for grub damage. Other signs of grub infestation are the presence of moles in the yard and frequent sightings of birds digging for grubs.

WATERING

Unless you are growing Bermuda grass or zoysia, which I don't encourage in this region, your lawn will naturally go dormant this month and return to green when the rains fall in September.

JULY

WHAT TO WATCH FOR

The biggest challenges to your lawn this month will be drought and heat. If you are growing buffalo grass, which is native to the prairie and the Great Plains, you should have no problems.

MOWING AND MAINTENANCE

Chances are your grass will be going into a light brown dormancy this month. That's actually good news, because you won't have to mow it as much. Don't worry, it will return to green this fall when the rains return.

FERTILIZING AND SOIL BUILDING

Do not add fertilizer this month. Do continue to make compost. Add kitchen scraps, especially all those good vegetable peelings, corn husks, melon rinds, spent flowers, and other good stuff you are now accumulating. Turning the pile every two weeks will really speed up the compost process, but you don't have to do it that often if you don't want to.

WEED CONTROL

Keep pulling out weeds from the lawn. Consider using Safer's Insecticidal Soap as an environmentally friendly spray to control any pernicious weeds that are causing you trouble.

PLANTING AND OVERSEEDING

August is a good time to plant seeds but not now. Buy your seeds, fertilizer, and organic matter and schedule a couple of work days to get the job done later.

PEST CONTROL

The beetles that were grubs are now laying their eggs, which will become grubs later. Avoid any light frequent watering of the lawn, because that will provide the beetles with the perfect egg-laying environment. It is better for the lawn to be dry, which helps dry out the eggs and kills many of them.

WATERING

Hold off any watering now; it only makes matters worse.

AUGUST

WHAT TO WATCH FOR

As the days get shorter and the nights get a little bit cooler, watch for your lawn to revitalize itself.

MOWING AND MAINTENANCE

Mid- to late August is a great time to plant new grass seed. If overseeding, lower the mower's cutting height and cut the grass as low to the ground as possible. Otherwise, continue your regular light and infrequent summer mowing pattern.

FERTILIZING AND SOIL BUILDING

If you are planting new grass seeds, you should definitely add a high-phosphorus natural organic fertilizer to stimulate root growth. Otherwise, hold off fertilizing until next month.

WEED CONTROL

Now is a good time to dig out weeds, because they, too, are weak and spent from a long summer of growing.

PLANTING AND OVERSEEDING

Six weeks before the first expected frost in your area is the very best time to plant grass seeds. That means from mid-August to late September for your region. Be sure to add organic matter, starter fertilizer, and a generous supply of expensive grass seed. Keep it watered and keep the kids off it until it starts to grow. Please read chapter three on lawn renovation for complete details.

PEST CONTROL

Now is a good time to apply grub-busting beneficial nematodes if you didn't apply any in the spring. The nematodes will die once the cold weather sets in, but you can kill many young grubs now, before they burrow deep into the soil for the winter.

WATERING

Avoid watering unless absolutely necessary.

SEPTEMBER

WHAT TO WATCH FOR

Your lawn will start looking green again as the rains return. Your trees will start to drop some of their leaves.

MOWING AND MAINTENANCE

Your grass is growing again, so keep on mowing. As a few of the first leaves start to fall, you should mow them right into the lawn, because leaves are a great source of phosphorus fertilizer, the type that makes your lawn's root system strong, and a great source of organic matter.

FERTILIZING AND SOIL BUILDING

Fall is the best time to apply natural organic fertilizer, because the grass plants are actually storing nourishment in their roots for the winter ahead, only to release that energy for a green spring lawn. Add an entire year's worth right now or the half remaining from your spring application.

This is also a good time to add gypsum to neutralize an alkaline soil or lime to neutralize an acid soil. Have a soil test done first.

This is a good season to top-dress your lawn with a thin layer of organic matter or compost to help build up microorganisms and prevent thatch.

You definitely need to empty your compost bin now, because leaf harvesting is soon to begin.

WEED CONTROL

Keep digging those weeds out by hand. You should have most of them out by now.

PLANTING AND OVERSEEDING

This is an excellent time to plant grass seed for a new lawn or overseed an old one.

WATERING

As the fall rains are beginning to return, you shouldn't have much need to water right now.

OCTOBER

WHAT TO WATCH FOR

When those colorful autumn leaves come tumbling down, you want to get out there and harvest them for compost and to keep them from smothering your grass over the winter.

MOWING AND MAINTENANCE

Put a bagging attachment on your mower and mow the leaves. The mower will shred them and collect them at the same time. Simply empty the bag into your compost bin or spread them around your flower beds for winter protection.

Lower your mower height to 2 inches when you are ready to cut your grass for the last time. This clips the grass low enough so that it doesn't fall over, mat down, and smother the turf over the winter, which can lead to diseases for the lawn next year.

FERTILIZING AND SOIL BUILDING

You can finish up your fertilizer applications this month if you live in one of the southern parts of this region.

PLANTING AND OVERSEEDING

There is probably not enough time to plant grass anymore this season.

NOVEMBER

WHAT TO WATCH FOR

Frost will be widespread now, and the ground will begin to freeze in the northern areas of this region. Be sure to gather up those leaves.

MOWING AND MAINTENANCE

Be sure to keep your cordless electric lawn mower plugged in and stored in a warm dry place for the winter. It keeps the battery working better longer. Be sure to let your gas motor run until it is out of gas and drain the oil before you put the mower away for the winter. Clean the mower really well now, especially the underside, where clumps may have built up that can damage the coating.

Avoid the rush next season and take your blades in for sharpening now.

FERTILIZING AND SOIL BUILDING

Hold off on fertilizing until next year. Keep making compost with shredded leaves, spent flowers, kitchen scraps, and other items.

WEED CONTROL

You can dig out weeds by hand until the ground is frozen.

DECEMBER

WHAT TO WATCH FOR

Santa's sleigh.

FERTILIZING AND SOIL BUILDING

Continue making compost. Turn the pile once a month until it freezes. Shred your Christmas tree and add it to the pile, or cut off its branches and lay them over your flower beds for winter protection.

Part Three

ENVIRONMENTAL

LANDSCAPING

Top Ten Ways
TO A
Beautiful Easy Landscape

Your home is not only the most important investment you will ever make, it is where you and your family sleep, eat, play, and sometimes work. Your yard is a valuable "green space" that surrounds your home. Typically, a yard has some grass growing, a tree or two, often a few shrubs, and maybe a bed of marigolds. But your yard can be many things:

- A beautiful landscape that is easy to maintain and enhances the value of your property by up to 18 percent.

- A charming and useful extra room where you, your family, and your friends share many good times.

- A wildlife habitat for birds, butterflies, and other animals, making it an educational classroom for you and your children.

- Part of the "green lungs" of your neighborhood, pumping in hundreds of gallons of fresh air, depending upon the types of trees, shrubs, and grass you grow.

- A primary component of an energy-efficient economy, reducing our dependence on volatile supplies of foreign oil and saving you and your country a lot of money. Trees and shrubs can provide shade and natural air conditioning.

Your yard can become all of these things and more if you plan to have a beautiful landscape that is easy to maintain without the use of chemicals. It really boils down to a matter of choices. You won't have to make any major, massive, expensive changes to your home landscape to achieve these results.

This chapter will tell you how to rethink your current landscape; to shrink the size of your lawn, going for quality not quantity; to create a backyard wildlife habitat that attracts birds, butterflies, and beneficial insects, by planting certain types of flowers and shrubs; to use food-bearing plants as landscaping tools and still have a beautiful ornamental yard.

You will learn how valuable trees are as a force to clean up our polluted air, and how the right trees planted in the right place can cut your heating and air-conditioning bills.

This chapter also has information on xeriscaping; water conservation; clean, easy compost; chemical-free disease and pest control; easy-care perennials, annuals, shrubs, and wildflowers; ground covers; and which plants are best suited for your area, all designed to make your yard beautiful to behold and a pleasure to maintain without the use of chemicals. Just follow the list of "Top Ten Ways to a Beautiful Easy Landscape" and you are well on your way to becoming an environmentally conscious gardener. Some of the material presented here may sound like a repetition of information given in Part One. In this chapter, however, I apply the ideas comprehensively to landscaping, not just to grass.

TIP NUMBER ONE: *Discover Your Ecological Landscape*

The first thing you need to do is survey your existing property to inventory what is already there, decide what you would like to do with your yard, and determine whether your yard and location can accommodate your goals. All of the information you gather here is going to be very useful to help you discover the potential of your ecological landscape.

1. What is your climate? What state or region of the country do you live in? This is obvious but important, because the type of ecoscaping you do in Vermont differs from the type you can do in Southern California. The temperatures, rainfall, winds, soil composition, and seasons are all different, and they limit your options and change your needs.

2. What is your microclimate? What is the immediate topography of your yard? Are you on top of a hill where it can be very windy, or in a valley where the winds are softer? Are you near a large body of water, a lake, a river, or an ocean? All of these can affect your ecological landscape.

3. Review the legal survey map of your property. There is no sense making plans and spending money on ecological landscaping without knowing exactly what your property boundaries are. Physically walk the perimeter of your property. Try to find the survey markers, if any, to verify your map.

4. Contact your state or local government zoning office about fence rights and responsibilities and whether any solar easement laws are in effect in your area. For instance, if you plant a large shady tree on your property, which also shades a solar water heater on your neighbor's property, your neighbor can force you to cut it down if he or she has solar easement rights. This is rare, but it does happen.

5. How many square feet or acres do you have on your property? How much of that is used for your house, your garage, your equipment shed or barn? Other uses?

6. How is the house situated? Does it face the street or road, and is that thorough-fare busy or quiet? What sorts of trees or shrubs are planted in your neighbors' yards? Do they hang over into your property, causing shade or an accumulation of yard waste?

7. Where does the sun rise and set over your house? Does the sun shine into the windows? Which way does the wind usually blow? Do trees and bushes growing on the south side shade the house in summer and winter?

8. What is growing in your yard? What types of trees, if any, and how big are they? Do they shed a lot of leaves? What sorts of shrubs and how big are they? Do you prune them often?

9. How much of your yard is planted in turf grass? Are there bare spots under trees or other problem areas?

10. Is your yard shady, sunny, or partly shady?

11. Does your yard have steep or slight slopes? Do you have a problem with water runoff after a heavy rain?

12. How do you use your yard now? As a football field or for entertaining, hanging around the swimming pool, patio parties, flower growing, hanging clothes on the line, growing vegetables, parking cars, or other uses?

13. How do you want to use it?

14. Do you now or have you ever used chemical pesticides, herbicides, or fungicides in your yard? Have you noticed any fireflies, birds, butterflies, bees, snakes, toads, or other small wildlife in your yard? What types?

If you have taken notes and answered all of the above questions, you are armed to start making decisions about the types of plants you can and should grow, whether your yard is large enough to accommodate all of your plans and activities, and what major hurdles you are going to have to overcome. Now that you have a complete catalogue of your yard, it is time to move on to the specific changes you can make to turn your little corner of the world into a Garden of Eden.

TIP NUMBER TWO: *Make Your Lawn Disappear*

I don't mean disappear totally. I mean it might be a good idea to reduce the size of your lawn and bring it back into balance with the rest of the great things you could be growing in your yard, and to reduce the amount of water, fertilizer, and pesticides it takes to keep that lawn looking clean and green. Most people take care of lawns that are just way too big. The old notion of a huge, perfectly mani-cured lawn planted fence row to fence row and taking up almost all the room you have in your yard is hopelessly out of step with today's dual family interests of quality leisure time and a cleaner environment.

Grass is a very hungry and thirsty part of your yard. It is so demanding that the Marin County, California, Water District is paying residents of the town of Novato, California, as much as $300 to rip the grass out of their lawns in their Cash for Grass rebate program. Homeowners have been using a third of Marin County's water supply to grow grass. This new program encourages them to switch to plants that don't require as much water as grass.

I am not suggesting that you rip out your whole lawn, although I have known people who have no grass at all in their yards, and it looks really great. I'm suggesting that you think of your lawn as a condiment rather than the main course, as just one of many acts rather than the main attraction, that your lawn become an accent to an overall lawn and garden plan. This is a two-step process, starting with size reduction and finishing with input reduction.

Size Reduction

Ground Covers. Ground covers are low-growing, rapidly spreading, relatively inexpensive, virtually maintenance-free plants that form a mostly weed-free carpet over the parts of the yard where you plant them. Once established, they require little or no watering and no pesticides. Ground covers are highly recommended for problem areas on your lawn such as at the base of trees or under shrubbery, in extremely shady areas, and on slopes, where they form a living erosion-control barrier. When planted under shrubbery, they form a living mulch, keeping mois-ture in the soil and deflecting sunlight, which keeps the soil much cooler. Some of the most common and widely available ground covers you can use are:

- ajuga, a.k.a. bugleweed
- English and other kinds of ivy
- Japanese spurge, a.k.a. pachysandra
- periwinkle, a.k.a. vinca minor and myrtle
- lily of the valley
- sedum, a.k.a. stonecrop.

Some other lovely ground covers include:
- wild European ginger
- sweet woodruff
- liriope
- bearberry
- dwarf astilbe.

Even hostas can form a great, slightly taller type of ground cover.

(Above, right) Make your lawn disappear—or at least shrink it down to a more manageable size. Ground covers such as ivy, sedum, pachysandra, and this myrtle and low-growing potentilla are naturally healthy and easy-to-care-for plants that you can grow as an alternative to lawns.

(Below, right) Some ground covers thrive in shade. Try the many types of hosta, for a medium-sized plant, and this low-growing lamium beacon silver.

Ground covers need only an average, not overly fertilized, well-prepared garden soil to be successful. Dig or rototill the area to a depth of 4 to 6 inches. Work in an inch or 2 of compost, dried manure, peat moss, or other organic matter. You can often buy ground covers in flats of ten to twenty plants from your local nursery or Soil and Water Conservation District. Plant the ground covers in a grid pattern, spaced evenly anywhere from 6 to 12 inches apart, to enable them to fill in the open spaces. Water to get them growing, and then forget about them.

Planting more than one type of ground cover in clusters next to each other creates a lovely woodland-type pattern. You can then plant small bulbs such as muscari, crocus, and dwarf varieties of daffodils and tulips in among the ground cover for spring blooms.

Wildflowers. Another good way to reduce the size of lawn you have to mow and maintain is to plant part of your lawn with wildflowers, bringing a section of meadow to your own backyard. Meadows in nature are never mowed but are eventually overgrown with shrubs and then trees. Your wildflower meadow will need to be mowed once a year, usually in the fall after all the plants have bloomed, but you can mow it in spring before the plants start to green up.

You don't have to turn your whole lawn into a meadow. Just pick an area along a fence or along the back of your property and plant it with wildflowers. Planting your front lawn with wildflowers can be very beautiful; one of my neighbors does it quite successfully with a succession of lovely blooms all summer long, but some towns may have ordinances prohibiting this practice on the grounds that it is unsightly.

Prepare a wildflower bed the same way you would a garden. Rototill or dig the soil thoroughly, incorporating a 2- to 3-inch layer of organic material. Smooth out the seed bed and scatter the seeds carefully. Rake them lightly to cover with $1/2$ inch of soil, and then water.

There are several wildflower seed mixtures on the market, often blended for open meadows, woodland gardens, and the like. Read the label carefully or ask the salesperson to help you. There must be an adequate mixture of annuals along with the perennials, or you will end up with no blooms the first couple of years while the perennials get established.

Edible Landscaping. This is the most delicious and nutritious part of your landscape program. Many people don't have the space or the time necessary to grow an extensive fruit and vegetable garden. But they still might want to pick a few berries or use fresh herbs in cooking and have a bed of attractive flowers, to boot.

Growing edible herbs, flowers, fruits, and vegetables as part of your landscape is a good way to have some homegrown food with very little work. It is really just a matter of substitutions. Here are some suggestions:

- *Strawberries.* Plant a row of strawberries along your front walk or driveway for attractive greenery all summer long. Low-growing alpine strawberries, a.k.a. fraises des bois, are particularly beautiful, and their tiny berries are very sweet.

- *Basil.* There must be twenty different types of basil plants available these days in garden centers and seed catalogues. These bushy plants have green or purple leaves that are large and flat or tiny and ruffled. They smell great, they are easy to grow, and they taste great in salads or chopped in pasta primavera. Even their flowers are lovely. Plant a few of each to give great texture contrasts to your bed.

- *Rosemary.* Rosemary grows rapidly from a small plant bought at the nursery, turning into hedges in some of the warmer parts of the country. Some people prune them to resemble trees. Break off a branch and toss it on the coals when barbecuing chicken to give off a wonderful Mediterranean fragrance.

- *Edible flowers.* Blossoms from calendula, violas, nasturtium, and chive plants are turning up in salads at fancy restaurants and even in salad packs at grocery stores. They are all very tasty, colorful, and fun.

- *Ornamental peppers.* These are compact, easy-growing bushes covered with red, orange, yellow, and green miniature peppers that can be very hot to eat.

- *Blueberry bushes.* Blueberry bushes can grow to the size of lilacs and produce several quarts of berries each year while still being ornamental and easy to grow.

- *Grapes.* Grow the vines on trellises as canopies for your back porch or gazebo. The vines will provide shade, and the delicious clusters hang down for the picking. This practice is common throughout the Mediterranean.

Input Reduction

I consider this to be a very important part of the lawn care equation because it means *less work* for you. Remember, this book is called *Beautiful EASY Lawns and Landscapes,* and easy is just as important as anything else we talk about. By simply reducing the size of your lawn, you are reducing the amount of time you have to spend taking care of it. But wait, there are even more time-saving tips that keep you off the mower and on to better things.

The established chemical lawn-care program of fertilizing four times a year, applying pesticides and herbicides, daily watering, bagging and dragging grass clippings to the landfill mounts up to an enormous amount of work and can be quite expensive. You can cut your expenses, cut your time spent, and cut the potential impact you are having on the environment. The chapters on lawn care earlier in this book give complete, in-depth information on environmental lawn care, but here are the basics in a nutshell:

1. Let your grass grow taller, up to 3 inches for northern grasses, St. Augustine and Bahia grass and 2 inches for Bermuda and zoysia grass. Tall grass better withstands drought and disease and controls a lot of weeds.

2. Leave the grass clippings on the lawn. This cuts your time spent by up to 40 percent and provides up to half of your fertilizer needs for free.

3. Plant the newer disease- and pest-resistant grass seeds that eliminate the need for pesticides.

4. Use natural organic fertilizer once a year, not four times as prescribed by some chemical companies.

5. Water only once a week during dry weather, or let the grass go dormant to return in the fall when the rains come back.

If you take pride in a good-looking lawn, this is the way to go. Your lawn will be smaller, easier to take care of. You will have a lawn that is better quality. It will be a green jewel, rather than a shaggy browned-out mess.

TIP NUMBER THREE: *Go Organic*

Let me reassure you that you can have a beautiful landscape that is easy to maintain without the use of chemicals. As I explained in Part One, a lot of people tell me they are, or a certain member of their family is, afraid to go organic because they have relied on chemicals for many years and found that that system of landscaping works just fine. But new concerns about the environment and family health have caused people to take a good look at the organic alternative.

Chemical landscaping depends on chemicals to control nature's sometimes less than perfect scheme to make the plants do what we want them to do. It is a top-down approach. We are in firm control, or so we like to think. Environmental landscaping, or ecoscaping, on the other hand, releases nature's own intricate system of checks and balances and lets the plants fight off diseases, pests, and other problems all by themselves. It is a ground-up approach.

Now, I am not a fanatic. There may be an occasional time when the judicious use of a relatively benign herbicide is a sensible option in the backyard landscape. It is the spray-first-and-ask-questions-later attitude that I object to and consider counterproductive.

Too many people forget that the landscape is a garden, too, just like their lawn and their flower beds. Whether your landscape is trees, shrubs, a wildflower meadow, or a bed of shade-loving hostas, you need to do four things to develop and perpetuate a beautiful easy landscape: 1. take a soil test; 2. use natural organic fertilizers; 3. replenish your soil with organic matter; and 4. make and use compost.

Soil Test. Soil has a chemical alkaline or acid balance that is measured on a pH scale of 1 to 14. The best garden soil for most landscape plants is a neutral soil, which has a pH of 6.0 to 7.0. Mountain laurel, blueberries, rhododendrons, and a few other acid-loving plants prefer a pH slightly below 6.0.

When your soil's pH is out of whack, your plants cannot chemically utilize the nutrition you offer them in fertilizers. They don't grow well, and weakened plants are the ones that become prone to pests and diseases.

Have your soil tested by your local Cooperative Extension agent; there is one in every county in the United States. The agent will tell you how much lime to add if your soil is too acid, or how much horticultural sulfur or gypsum to add if your soil is too alkaline. Recent studies have shown that generous additions of compost, especially homemade compost, go a long way to balancing soil pH all by itself. Peat moss will also naturally acidify an alkaline soil.

Organic Matter. The importance of adding organic matter to your landscape cannot be overstated. A soil rich in organic matter will encourage plants to develop a better root system, which loosens the soil and makes it more water retentive, which allows more air to get to the soil, which provides the food source, oxygen, and moisture for microorganisms, worms, fungi, and other beneficial animals to thrive in your soil, which will help fight off most diseases and insects.

A soil rich in organic matter is the absolute foundation of a beautiful easy landscape. It is the only way you can have a beautiful landscape that is easy to maintain without the use of chemicals.

The most accessible forms of organic matter are compost, bagged humus, shredded leaves, grass clippings, crop residue, old hay, composted manure, and peat moss. Whatever kind of natural organic mulch you use is also a form of organic matter.

Organic matter is perishable. The microorganisms in your soil are constantly digesting, using up your supply of organic matter all the time. You have to add more every year. Here's how:

- Spread a layer of organic matter over your perennial flower beds each fall when you put the garden to bed for the winter.

- Spread a layer of organic matter over your annual flower or vegetable beds each fall and till or dig it into the soil. By spring it will all be decomposed.

- Whenever you plant a tree or a shrub, dig a shovelful of organic matter into the soil as you return it to the hole.

- Use organic matter such as compost or shredded leaves as part of the annual mulch that you spread around the garden or under the shrubs.

Compost. The very best organic matter you can use is compost. And it is free. Last spring when I emptied my compost bin and spread the humus around my flower beds, I calculated that if I were to purchase an equivalent amount of bagged humus or composted manure, it would have cost me $50 to $75. For a complete discussion of compost, please read the chapter "Clean, Easy Compost" earlier in this book, but here is a nutshell summary:

1. Buy a sturdy, hard-plastic, wooden, or wire compost bin with a tight-fitting lid. It is important that your bin look good so your neighbors don't complain. Home-made bins are fine if you live in the country, but be warned! A flimsy bin in the country means certain critter attacks and compost strewn all over your yard.

2. Place the bin in a sunny or semi-shady place on level ground that is not too far from your house. Behind or beside the garage or in the back by the alley is the usual spot.

3. Fill the bin with shredded leaves, uncoated paper plates, paper towels, paper napkins, nonanimal-source kitchen scraps, crop residue like over-the-hill flower buds, and any other type of organic matter you find around your landscape. It is best if you can layer this material, but it is more important that you keep adding things.

4. Turn the material once a month with an easy-to-use compost aerator. Toss in a bucket or two of water if it looks like it is drying out. In six months or so, you will have a bin full of stuff that looks more like crumbly brown soil than banana peels. Stop adding material. Turn the material once a week for one month. Voila! Compost!

5. Empty the bin and spread the humus around your landscape. Start all over again. Repeat process. Voila! Compost!

Natural Organic Fertilizers. Most people have been buying synthetic chemical fertilizers over the years because that's what was widely and inexpensively available at the garden center. Besides, they do give you big plump tomatoes and large showy flowers. But natural organic fertilizers are a better alternative that will also give you great produce and lovely blossoms.

Natural organic fertilizers are granular or powdery; they are dry, and they don't smell any more than chemical fertilizers. They are made from agricultural by-products such as bone meal, blood meal, ground grains like wheat germ and cotton seed, composted manures, and naturally occurring minerals.

One of the benefits of natural organic fertilizers is that they are slow acting. This means there is no chance they will burn your plants, and, furthermore, they will feed your plants all season long with only one application. Finally, whenever you use natural organic fertilizers, you will be adding organic matter to the soil.

Most stores now carry natural organic fertilizers. If you have any questions, just look at the label or ask the salesperson to help you.

TIP NUMBER FOUR: *Make Every Little Drop Count*

Everyone knows by now that water is a precious commodity. As our population grows, demand for water becomes even more intense. And fresh water is expensive. You have to pay big taxes to build megabuck water treatment facilities or huge reservoirs that seem to be inadequate as soon as they are built.

We know that nature doesn't always cooperate. In the summer of 1993, when the Midwest was deluged with floods, crops were drying out in the Carolinas and New England. What's a person to do? Well, you could do what my friend's neighbor does. He waters his lawn and landscape on a timed watering system, and he has great-looking grass. But his water bill for July was over $300. That's a lot of money. Here are some other, more environmentally harmonious alternatives:

Build the soil. Both clay soil and sandy soil have problems retaining water. Clay soil will hold a lot of water but not if it is hard and dry. Sandy soil just does not hold water very well at all. By adding a 2-inch layer or more of organic matter to your beds each year, you will loosen up clay soil and tighten up sandy soil. You will create a soil that holds water, that will soak up water like a sponge, that will store it like a vast underground reservoir and keep it there for the plants to use during drought periods. As I detailed above, adding organic matter to the soil is as simple as spreading it over the top of your perennial beds or digging it into your annual beds or spreading it under and around your trees and shrubs.

Install soaker hoses. Soaker hoses use up to 70 percent less water than other forms of watering your landscape. They are made of a type of rubber that allows water to ooze out the pores all along the length of the hose. You simply install your soaker by snaking it around in your beds, cover it with mulch, and leave it there. In most climates it can be left out year-round. It can even be buried a few inches and still ooze out water.

The reason soakers are so good is that they deliver water to the soil very slowly, so that it can be immediately absorbed in the soil. It also delivers water right at the soil level, which means very little is lost to evaporation as happens with other types of sprinkler systems.

You don't hook your soaker hose or hoses up to the spigot. Instead, run your regular hose out to the beds and hook it up to the soaker hose installed out there. Instead of watering for twenty-minute intervals three times a week, you turn on

A soaker hose delivers water to your flowers and shrubs in the most economical and effective way possible. The water seeps out of the pores of the hose very slowly and gradually percolates down deep into the soil. Very little precious water is lost to evaporation, and there is no chance of the leaves getting wet, which can promote disease on some plants.

the soaker hose once a week for one or two hours, depending on how dry things are. That saves time and money right there.

Cover your beds with organic mulch. Now your soil is a perfect sponge and reservoir and your soaker hose is there to supplement your rainfall amounts. But you still need one more water-conserving tool: that is natural organic mulch. Mulch creates a barrier that allows rainwater to drip down into the soil but then traps the water in there to prevent evaporation. It also reflects heat from the sun, keeping the soil beneath cooler, a condition plants like.

Black plastic mulch does a good job of holding down weeds and keeping moisture in the soil, but it is unsightly, it does not allow for rainwater penetration, and it does not decompose and add organic mulch to the soil. Breathable landscape fabrics are a good alternative because they stifle weeds and allow rainwater to penetrate, but they don't decompose either.

I like to use natural organic mulches because they stifle weeds, allow rainwater to penetrate, and decompose to add organic matter to the soil. You can use whatever organic mulch is popular in your area, from pine bark to cocoa bean hulls to straw. I use a combination of homemade compost, shredded cedar or pine bark, and a little bit of peat moss. I find this to be economical and attractive, and it does a good job.

Whatever mulch you use, simply place a 2- to 4- to even 6-inch layer around your flowers, herbs, shrubs, and trees. You will have to replenish it each year, but you probably will not have to use as much after the first year.

Install a rain barrel. Rain barrels attached to downspouts to collect water have been in use for thousands of years. I have lived in two homes, one in Spain and one in Boone County, Missouri, that had a cistern, which is really an underground rain barrel, as the only water supply.

As seriously discussed, rain barrels come in all shapes and sizes; you can pick one that fits your aesthetic needs. You might consider a large 20-gallon washtub, or a wooden full or half barrel, or you might look at a 50-gallon hard-plastic rain barrel that really holds a lot of water. Here are three good catalogue sources for rain barrels:

Great American Rain Barrel Company, 295 Maverick Street, East Boston, MA 02128, (617) 569–3690.

Gardener's Supply, 128 Intervale Road, Burlington, VT 05401, (800) 863–1700.

Gardener's Eden, P.O. Box 7307, San Francisco, CA 94120, (800) 822–9600.

You simply place the barrel under your downspout and start collecting water whenever it rains. You can also install a rain barrel to the downspout of your garage, which will give you less but still plenty of water. I have one installed on a garage downspout next to my herb garden, and I find it really useful. Many of the larger plastic rain barrels also have spigot attachments that you can hook a hose up to, and connecting tubes that allow you to fill additional rain barrels with the overflow from the first one. Once you have water in the barrel, just dip your watering can in the barrel and fill it up. It's all very old-fashioned, but it's also the modern cutting edge of water conservation.

Xeriscaping. Xeriscaping is a term that is derived from the Greek word *xeros*, which means "dry." But that does not mean desert or deserted. Too many people think of rocks, gravel, sagebrush, and cactus when they hear the term xeriscaping.

Xeriscaping really means gardening and landscaping with water conservation in mind. Although xeriscaping has become more accepted in the parched areas of Southern California and Texas, where water is often in short supply, it is becoming better known as it is adapted to south Florida, where too many people and too little water have created severe water shortages, as well as the central Midwest, where several years of drought killed off a lot of lawns.

Choose drought-resistant plants that grow well in your area. It is impossible to list all the different plants for all the different areas of the country. It is best to consult with your local Cooperative Extension agent or garden center for advice.

TIP NUMBER FIVE: *Bring Back Birds, Bees, Butterflies, and Bats*

Birds, bees, butterflies, and bats are four of the most beneficial creatures you can have in your backyard landscape. Besides each being beautiful in its own way, they all make life easier and more pleasant. For instance, each harmless little bat you see darting around is devouring nearly one thousand mosquitos each night. Birds also eat mosquitos and scads of other damaging bugs. Bees pollinate flowers, giving your display more bloom. Butterflies are great pollinators, and they are also just plain beautiful.

Wildlife has come under increasing pressure for the last fifty years because of excessive and indiscriminate spraying of chemicals and the loss of habitat. The building of new housing developments and subdivisions has meant the destruction of meadows, abandoned farms and fields, woodlots, stream banks, and brushy areas that provide food and shelter for animals. You can help turn the tide by providing wildlife a healthy place to live and thrive right in your own backyard. Here's how:

Attracting Birds

Birds are the most rewarding and most beneficial wildlife you can attract. Birds are beautiful and fun to watch, and they sing lovely songs. They also eat enormous amounts of bugs like mosquitos, which will make your life outdoors more pleasant.

The first thing you have to do to attract birds is to stop spraying chemical pesticides on your lawn and garden. Thousands of birds are killed each year by pesticides.

Next, create "edges" along the sides of your property. Edges are the areas where trees meet shrubs, where shrubs meet flowers, where flowers meet the lawn. Birds love edges because they provide cover and food.

Edges should also have layers—in other words, taller trees mixed with shorter trees and bushes and varying heights of flowers. Different birds thrive at different heights. To ecoscape for edges and layers in your yard, think of a hedgerow on an old farm. There is usually a long row of different types and sizes of trees, often covered with different types of vines and dead branches hanging down. Beneath the trees are shrubs and bushes fol-

Attract birds to your yard as a natural form of insect control. Many birds feast on mosquitoes, beetles, grubs, and other insect pests. Contact your local Audubon Society, lawn-and-garden center, bird feed store, or Cooperative Extension office for advice on choosing the right kind of bird feeder, bird food, and birdhouses for the birds in your area.

lowed by flowers and brush growing close to the ground. Finally, there is the meadow or fallow field. This is ideal habitat for most birds. The more varieties of edges and layers you create in your landscape, the greater number and variety of birds you will attract.

Plant vegetation that provides food for the birds. Instead of forsythias and lilacs, plant flowering dogwood, hawthorn, holly, or euonymous, all of which have fruit for birds. Birds also love to feed on red chokeberry, barberry, mountain ash, winter berry, rosa rugosa flowering rose hedge, crabapple, and firethorn.

Naturally, you can always feed birds from a bird feeder, but remember that birds need perches and nesting spots to thrive, not just birdseed. You can feed birds year-round, and it will not make them lazy and cut down on their insect eating.

Birds also need water for drinking and bathing. Set out containers of water or a birdbath or even build a small pond or pool for them. They prefer running water; if you can build a small water garden with a fountain, birds will love you for it. Be sure to change the water at least once a week for a stationary birdbath.

Flowers, especially those from the sunflower family, can also attract birds. They like to eat the seeds in the flower heads in the fall from these varieties: aster, bachelor's buttons, campanula, chrysanthemum, cornflowers, cosmos, marigolds, phlox, verbena, and zinnia. These flowers are all very easy to grow from seed in a sunny, well-drained soil. Of course, anyone who has tried to grow cherries, blueberries, and other berries knows that birds always know when those fruits are ripe.

Attracting Hummingbirds

Hummingbirds are good for the garden because they eat insects and help pollinate flowering plants. Of course, they are also exciting to watch.

Attract hummingbirds by planting flowers, shrubs, trees, and vines with tubular blossoms. Some of their favorites are trumpet vine and trumpet honeysuckle, salvia, fuchsia, hollyhock, morning glory, four o'clocks, nasturtium, larkspur, and mimosa tree. Hummingbirds are known to be attracted to the color red in flowers, but I have seen them feast on the nectar of many colors of plants.

Attracting Butterflies

Attracting butterflies is as simple as planting flowers with purple, yellow, white, and pink blossoms. Butterflies need two types of flowering plants to survive. They will light on a wide variety of big colorful ones with nectar to provide them with food, and they need a few more specific ones to lay eggs on. In general, many butterflies prefer to lay their eggs in more secluded, undeveloped areas on uncultivated plants such as the umbelliferous Queen Anne's lace.

Since you probably don't want to attract leaf-eating caterpillars to your gar-

Attract butterflies, hummingbirds, and bees to your yard by planting a variety of colorful annual and perennial flowers that bloom at different times during the season. Butterflies are in desperate need of food and habitat because of mosquito abatement spraying programs. Hummingbirds and bees help pollinate all sorts of flowers, making everyone's garden more fruitful and beautiful.

den, even if they are going to be butterflies some day, you would do better planting the flowers they use for food. Some good plants are butterfly weed, butterfly bush, cosmos, zinnia, asters, daisies, mock orange, morning glories, primrose, sweet William, gaillardia, scabiosa, phlox, and sedum. The crucial point is to have flowers coming in to bloom over the entire course of the growing season, so the butterflies have nectar all season.

Attracting Bats

Bats are our friends. They are not evil vampires, they do not get caught in people's hair, and they do not want to attack us. They want to fly around at night and eat insects and then they want to be left alone. You really don't want millions of bats living in your attic, but I have a few and I am going to leave them there because a single bat can eat nearly one thousand mosquitos each night.

If you have bats living in your landscape, leave them alone. If you want to attract them, install a bat house, which is very similar to a bird house and can be bought at many garden centers or through several garden catalogues.

For more information about bats contact Bat Conservation International, P.O. Box 162603, Austin, TX 78716; (512) 327–9721.

Spiders, bees and wasps, snakes, and toads are also good creatures to attract to your yard, but not everyone wants these animals around. Spiders eat insects; bees and wasps help pollinate fruits and vegetables, making them more productive; snakes eat rodents; and toads eat a huge number and variety of insects.

Get to know more about your wildlife friends. There are many nature centers and wildlife preservation groups that would be happy to supply more information. You may also want to read *The Naturalist's Garden* by Ruth Shaw Ernst (Globe Pequot).

TIP NUMBER SIX: *Create an Urban Forest*

Managing existing trees or planting new ones is the most influential and far-reaching step you are going to take in your backyard landscape program. Planting the right kinds of trees in the right place on your property can give you three big environmental pluses: clean air, energy savings, and beauty.

Plant a tree and clear the air. The magic of trees is that they change dirty, polluted air into clean air. They photosynthesize CO_2, carbon dioxide, the prime component of global warming and the greenhouse effect, into O_2, oxygen, which we humans need to breathe. Scientists say we have a symbiotic relationship with trees. In other words, we need each other to survive.

The United States is only 4.8 percent of the world's population, yet we produce 25 percent of the global CO_2. A single mature tree growing in a rural area consumes 13 pounds of CO_2 each year. That same tree in an urban area will consume fifteen times that amount, because the CO_2 level in cities is so much higher.

Save energy and save money. The National Arbor Day Foundation estimates that shade trees can save up to 50 percent of air-conditioning costs for your home in the summer. Evergreen trees as windbreaks can reduce heating bills by as much as 30 percent in winter.

The American Forestry Association estimates that one shade-providing fifty-year-old tree will save the average American household $73 per year. The same tree will save the same household $75 per year in costs associated with soil erosion and storm water control. (Cities have to build adequate storm drains for water control during heavy rains. If there was less runoff, because of water retention provided by trees, they would have to build smaller storm drain systems.)

Add another $50 a year for air pollution control, plus other cost savings such as wildlife shelter, and the total yearly savings for growing and nurturing a single mature tree on your property will be approximately $273.

Trees are beautiful. Finally there are the intangibles that you get from planting trees. Trees offer you a sense of history, a sense of place, a sense of belonging. In our fast-paced mobile society, these factors can help relieve stress.

Many trees bloom in the spring, notably tulip, cherry, dogwood, locust, and hickory. Their fragrance and beauty bring charm and a feeling of comfort to your home and neighborhood.

A Crash Course in Urban Forestry

Your property, your block, your neighborhood, and your town are all a part of a massive urban forest, which includes tree-lined streets, parks, roadways, cemeteries, public gardens, corporate grounds, and other public tree-bearing areas. Just fly over your neighborhood in a plane, and you will see what I mean. Many grandparents fondly recall the beauty of their streets canopied by giant American elm trees in the years before the Dutch elm blight. Tree-lined streets provide shade, privacy, noise control, natural air cleaning, and beauty. It is a tradition straight out of Norman Rockwell.

That vision is fading in the United States, because trees are dying and being cut down faster than they are being replaced. An average of only one tree is now being replanted for every four that die or are removed each year. Furthermore, an additional 40 percent of all trees are in a state of decline, with only a decade or two of life.

The American Forestry Association has started a program called Global Releaf to stem the tide of fallen trees. It wants to encourage the planting of two hundred million new trees by the year 1994. It estimates that one hundred millon new trees would reduce CO_2 emissions by eighteen million tons per year and save $4 billion in energy costs each year by providing shade and windbreaks. For more information contact Global Releaf, The American Forestry Association, P.O. Box 2000, Washington, DC 20013, (202) 667–3300.

The National Arbor Day Foundation also has tree-planting programs, including:

Trees for America. This program will send you live trees if you become a member. For a $15 annual membership contribution, for instance, they will send you ten pine trees suitable for planting.

Country Living Magazine National Arbor Day Forest. For a $15 contribution, this program will plant ten trees in your name in Yellowstone Park and also send you ten trees to plant on your own.

Tree City USA. This program helps towns and cities establish tree boards, set aside money to plant trees in urban areas, and celebrate Arbor Day.

For more information, contact the National Arbor Day Foundation, 100 Arbor Drive, Nebraska City, NE 68410, (402) 474–5655.

The Nuts and Bolts of Choosing and Planting Trees

Most trees prefer to be planted in the early fall when the ground is still warm and moist from autumn rain. This gives the tree time to establish its roots before the cold winter sets in. But some areas are too cold for fall planting. Consult with your nursery or county agent for what is best in your area. If you plant in the summer or spring, be sure to water your tree generously and often to keep it alive the first year. After that, watering should not be necessary except in cases of extreme drought and then only for one- to two-year-old trees.

1. Choose a tree that is right for your needs. If you want shade for your house, try a hard wood such as maple, oak, walnut, poplar, or ash. If you need a tree that can line a heavily trafficked street with a lot of auto exhaust, try a ginkgo or sycamore. If you want a smaller decorative tree, try a redbud or dogwood. Ask your local nursery for advice on what trees grow best in your area.

2. Give your tree plenty of room. Too often homeowners plant trees too closely, turning their yard into a jungle. Your tree needs plenty of room to grow and send out roots.

3. Dig a hole at least twice the size of the root ball on your tree. Loosen the soil and an additional 6 to 8 inches in the bottom of the hole to give the roots a better surface to dig into. Be sure the hole is dug squarely and not like a saucer.

4. Remove any wrapping around the roots and spread out the roots. Place the tree in the hole and spread out the roots again. Set the tree in the ground at about the same depth as it was when it was in the ground at the nursery. Fill the hole with crumbled dirt mixed with good compost. Tamp the dirt firmly around the roots.

5. Place a 2-inch layer of compost or peat moss on top of the planting. Water thoroughly. Do not fertilize the first year.

TIP NUMBER SEVEN: *Landscape for Energy Savings*

I know this sounds a lot like that pie-in-the-sky solar energy stuff that was popular in the 1970s during one of our many energy crises. But using energy wisely just makes good sense. If we as gardeners can do our part and still have the beauty we want in our landscapes, so much the better. Besides, I haven't seen my energy bill come down too much recently, have you?

The keys to landscaping for energy conservation and taking advantage of the sun's radiation are to create shade in summer, allow sun in winter, channel cooling winds in summer, and block cold winds in winter. You can accomplish most of those things with thoughtful plantings of trees, shrubs, and vines, plus a minimum of carpentry.

The two climatic elements you are trying to control are the sun, which gives off desirable heat in winter and unwanted heat in summer, and the wind, which blows cold and nasty in winter and hot and nasty in summer. You need to pinpoint two things: which way south is from your house, and which way the prevailing winds blow in your area in both winter and summer.

Planting for Sun and Shade

The sun may rise in the east and set in the west, but it always travels across the southern sky. On the longest day of the year, June 21, the first day of summer, the sun is directly overhead. From then on, it progressively dips farther south as it approaches the first day of winter, December 22, the shortest day of the year. The higher the sun is in the sky, the more shade you will need. The lower the sun is in the sky, the more access to the sun's rays you will need. In summer you want to block the sun. In winter you want to open your house to the sun.

Old-time contractors in the seventeenth and eighteenth centuries usually tried to face a house to the south. In other words, build the house with the largest side of the house facing south. That was usually the front but not always. On a clear sunny day the sunshine on your house will naturally heat the areas that receive the most exposure to the sun's light. What you want to do is utilize that heat in winter by not blocking those rays with trees and shrubs.

You need to locate the south side of your house and see what trees and shrubs are growing there; what sorts of trellises, garages, decks, or overhangs have been built there; and, most important, what type and size of windows have been installed there.

Under ideal conditions, you will have one or two large deciduous trees planted on the south side. Their leaves block the sun in summer, and they drop their leaves for solar penetration in winter. Remove any evergreens, shrubs, or trellises that block the south side of the house. Instead, plant two or three deciduous trees. What type of trees? Some experts suggest thin-branched trees such as ginkgo or Kentucky coffee bean tree, but these trees are too messy, dropping their nuts all over your lawn. It might be better to plant hardwoods such as maples, and oaks, which will grow tall and dense and fully shade the roof of your house for maximum summer cooling.

To shade a one-story home about 20 feet high, plant the trees 15 to 20 feet from the house. As these trees grow larger and their branches spread out, they will lower indoor air temperature in summer, reducing your need for air conditioning, and increase indoor air temperature in winter, reducing your need for heating.

Planting vines on trellises along the side of your house can also be a good way to create shade and help cool the house in summer. Often you don't have room to plant a large tree. Vines are a good alternative.

Deciduous vines, those that grow their leaves in summer and drop their leaves in winter, are the best bet. Boston ivy, wisteria, Virginia creeper, and clematis are all good vines to grow for shade. It is best not to grow the vines directly on the house, because they might damage the wood. Instead, grow the vines on trellises placed no more than 12 inches from the wall.

Planting for Wind and Windbreaks

You can cut your winter heating bills by 10 to 25 percent by the use of windbreaks if you live in the part of the country with cold winters. Windbreaks work by reducing air flow around the home, which cuts down on heat loss from the walls of the building. Windbreaks are important in the hot southern and southwestern parts of the United States, but they are necessary in the colder temperate zones in the North.

Since the prevailing winds in the colder regions of the United States are from a northerly and northwesterly direction, you want to plant your windbreak on the north and northwest side of your property. Evergreens such as pines and spruces make the best windbreaks. Give the trees plenty of room. If you expect the tree to grow 20 feet tall, plant it 15 to 20 feet away from the house. Space the plantings 8 to 10 feet apart.

The U.S. Department of Energy has divided the United States into four climatic regions. Each zone has landscaping needs that can help you reduce your fuel bills in both summer and winter.

The Cool Region. Northern tier of states from Wisconsin west to Montana plus Maine. Summers are warm, and winters are very cold. Winter is the problem here, and you need to focus on using plantings to insulate, deflect cold winds, and collect as much warmth from the sun as possible. Be sure no trees or hedges block the sun's path in winter. Plant a strong row of evergreens for a windbreak. Plant evergreen hedges such as rhododendron, juniper, or hemlock around the foundation of the building to create insulation.

Best deciduous trees include silver maple, chestnut, birch, oak, and crabapple.

Best evergreen trees include balsam fir, spruce, Scotch pine, hemlock, and Douglas fir.

The Temperate Region. From New England south to Virginia, west to San Francisco and north to Washington. Summers are hot, and winters are cold. Here you need to block the sun and gather breezes in summer, and block the wind and gather the sun in winter. Plant coniferous windbreaks on the north and west corners, tall deciduous trees on the south side.

Best deciduous trees include Norway maple, hickory, dogwood, hawthorn, and magnolia.

Best evergreen trees include Norway spruce, blue atlas cedar, Japanese plum yew, Canada hemlock, and Scotch pine.

The Hot and Humid Region. The Southeast from North Carolina to East Texas. Warm winters and very hot and humid summers. The object here is to create shade during the summer that also allows for summer breezes to cool the house. Plant shorter trees and shrubs along the east and west and taller trees for shade on the south. Plant ground covers around the home to absorb heat and cut down on reflection.

Best deciduous trees include palm, flowering dogwood, myrtle, tulip tree, and pin oak.

Best evergreen trees include Yulan magnolia, weeping fig, live oak, camphor tree, and citrus trees.

The Hot and Arid Region. The Southwest from Dallas to Los Angeles. Dryness and temperature extremes are the problems here. You need to plant for shade but also for moisture retention. Tall trees and shorter shrubs planted around the house create an oasis effect offering cooling shade and moderate temperatures.

Best deciduous trees include ginkgo, California maple, crabapple, almond, and Chinese pistachio.

Best evergreen trees include carob tree, desert gum, live oak, common olive, and Joshua tree.

For more information on energy-saving landscaping, you may want to read *Landscaping that Saves Energy Dollars* by Ruth S. Foster (Globe Pequot).

TIP NUMBER EIGHT: *Plant Easy Care Flowers, Shrubs, and Vines*

One of the delights of a beautiful easy landscape is the chance to enjoy all the different colors, shapes, and fragrances of blooming plants. Unlike vegetables, flowers demand very little time and fertilizer, and they don't seem to be as interesting to pests, either. Flowers are basically very easy to grow, once you get to know them a little bit, and if you choose the right ones to plant.

Perennials, those flowers that come up every year on their own, are, of course, by definition easy to grow. Once you get them in the ground, most of the work is over.

Annuals have to be planted every year, but most of them are very easy to get started from seed right in the garden, and once they get going, they provide dashing color and foliage all season long. Shrubs are the main anchors of a landscape after you have decided which trees to plant. I am not a fan of hedges that are closely cropped and shaved off into blocky boxlike shapes. I like my hedges to grow a little on the wild side and prune them with shears rather than manicure them with clippers. I am listing a few here that are colorful and don't require much work.

Vines are also a wonderful edition to any landscape. They are easy to grow, easy to take care of—especially if you like a slightly natural look—and they have lovely fragrant blossoms.

Now, let me just say that the following lists are by no means comprehensive. There will be a lot of more advanced gardeners and horticulturists who laugh at my suggestions and consider them just downright common. But I don't care. I'm not a horticulturist. I am a backyard gardener, just like you, who is learning as I go along. I think this is a great list for you to choose from to get started, and all of these plants will be easy for you to get your hands on at the local garden center or through easily available catalogues. Be sure to cross reference these lists with the ones for energy-wise landscaping (pages 151–154) and plantings to attract birds, butterflies, and hummingbirds (pages 145-149). Believe me, I want to learn more, too. If you have any suggestions, please send them along to me at P.O. Box 354, Claverack, NY 12513.

Perennials

Achillea. A.k.a. yarrow. This is a very drought-resistant plant that grows 3 to 4 feet tall in full sun to part shade. It has fernlike leaves and a many-colored blossom resembling Queen Anne's lace.

Aster. These are a tamed version of their wild cousins growing on roadsides. They grow 3 to 4 feet tall with a mum- or daisy-like bloom in purple, pink, or white.

Astilbe. Lovely fernlike plants that send up pink or white spiky blooms in mid- to late summer. One of the better shade plants.

Bee Balm. A.k.a. monarda or bergamot. Bees, butterflies, and hummingbirds love this red, pink, or white flowering 3-foot tall plant. It spreads like crazy and seems to withstand all types of neglect.

Bellflower. A.k.a. campanula. These are delicate-looking but very hardy plants from 8 inches tall to 3 feet in either blue or white flowers.

Bleeding Heart. A.k.a. dicentra. This is the exotic-looking spring-blooming foundation of any shade garden. It will bloom all summer long if kept well watered.

Chrysanthemum. There are dozens of different colors and shapes to the flowers of chrysanthemum, but they are all very hardy and love to spread out for a lovely fall bloom.

Coral Bells. These have lovely low-growing leaves that shoot forth with tall spiky yet delicate blossoms. They will grow in partial shade or full sun.

Coreopsis. One of my favorites, this yellow- and sometimes pink-blooming 2-foot-tall plant blooms all summer long and attracts butterflies with the best of them.

Daisies. Hardy, indestructible daisies come in many different colors but mostly similar flower shapes. I'm taking some liberties here and including yellow black-eyed Susan, white shasta daisy, golden gloriosa daisy, and purple cone-flower. Mix them all together for a beautiful 3-foot-tall hedge of summer-long color.

Day Lily. A.k.a. hemerocallis. Day lilies signal the beginning of the warm summer season where I live in the Hudson Valley. But familiar orange day lilies are only part of the story. They are also available in yellow, pink, peach, red, and many more colors.

Ferns. There must be twenty different types of ferns available that will grow easily in shady areas of your garden and fill the landscape with showy green.

Geranium. This is not the bright red potted geranium we are used to. This is a perennial geranium that forms a nice mat of foliage covered with blossoms all summer long.

Hosta. Hosta are green-fleshed leafy plants that do well in full shade or part sun. They spread like crazy, never seem to want much water or fertilizer, and stay green all summer long. In July they send up exotic-looking blossoms on tall spikes, which soon die out.

Lily. Even though they look delicate, lilies are actually quite hardy, resist diseases and pests, and require little in the way of fertilizer or water. I've mowed over lots of lilies in my yard, and they keep coming back. You can even get reasonably priced bulk lilies that can be naturalized like daffodils.

Lily of the Valley. Twelve-inch-high lily of the valley grows in deep shade, reliably sends forth its delicate fragrant white flowers in early spring, and then remains as a lovely green ground cover for most of the summer.

Peony. Peonies are an old-fashioned flower that grows on a large shrublike bush, which dies back each winter. Even though their blooming time is short and in early summer, there is really no more beautiful blossom than the peony.

Annuals and Bulbs

Cleome. A.k.a. spider plants. These are tall, rangy 4- or 5-foot-tall plants with big spiky blooms. They seem to be indestructible and fill huge areas in a garden. Easy to grow from seed.

Cosmos. Cosmos has lovely pink, blue, white, or red blossoms set on branchy stems covered with graceful foliage. They are tall and colorful and easy to grow from seed.

Dahlia. Dahlias are grown from bulbs that you plant in the spring and take out of the ground in the fall. They can grow from 12 inches tall to 3 feet with all sorts of colors and textures to their blossoms.

Four O'Clocks. These are 2- to 3-foot-tall bushy green plants with multicolored blossoms. They aren't show stoppers by any means, but they are incredibly easy to grow, and they take up a lot of space. I planted a packet of four o'clocks along the side of my house when I was a bachelor living in Missouri, and they bloomed all summer long. Now that's easy.

Gladioli. Gladioli corms are also planted in the spring and removed in the fall. The foliage is composed of long green, flat stems surrounding exotic colorful blossoms. One of the best cut flowers.

Iris. Iris look a lot like gladiolus, but they bloom in early summer from tubers you plant very close to the surface. The only work you have to do is cut the foliage back after the blooms have spent to prevent iris borers. That's it. They are very hardy and don't have to be dug up except to give them away to friends.

Marigold. I like marigolds. They are spunky, fragrant, and keep blooming all the way to hard frost. Easy to grow from seeds, I prefer the small French-style marigolds.

Nasturtium. Nasturtium is a low-spreading leafy green plant that sends forth beautiful blossoms of red, pink, and orange. They have seeds the size of peas, and they can be planted by simply poking the seed into the ground in any bare spot your garden has revealed.

Shrubs

Beauty Bush. This is a 10-foot-high, 6-feet-in-diameter, green bush that cascades with pink flowers in early summer right after the lilacs have bloomed. Very easy to grow.

Butterfly Bush. Buddleia is the other name for this tall spreading bush that sends out lilaclike pink blossoms all summer long. Butterflies love it.

Holly. Holly is an easy-to-grow plant that likes acid soil and some shade. It produces red berries that birds like to eat and green foliage in winter that makes great wreaths.

Lilac. If lilacs weren't so easy to grow, there wouldn't be so many of them in old farmyards and along roadsides. They need only a light pruning each year and then a deep pruning once in a while to remove dead wood.

Mock Orange. Mock orange is another fast-growing bush that takes up a lot of room in a short period of time. They have lovely orange, fragrant blossoms in early summer. You really don't have to do anything to them.

Mountain Laurel. There are few things as beautiful as wild mountain laurel in bloom along the roadsides in Connecticut and western Massachusetts in midsummer. These bushes need only a little bit of acid soil and little else to produce their red, pink, or white rhododendronlike blossoms year after year.

Spirea. Spirea is making a comeback as people start to recreate English cottage gardens. They also are indestructible, they fill large spaces in a very short period of time, and they flower exuberantly with white blossoms in early summer and stay green the rest of the season.

Vines

Clematis. Clematis, a perennial, is the bougainvillea of the northern garden. It is a vigorous grower. It will withstand quite a bit of neglect and send out beautiful flowers year after year.

Honeysuckle. I love the fragrance of honeysuckle, and I don't mind all the bees it attracts. The vine grows well along fences and takes the sharp edge off a chain-link fence. Honeysucke is a perennial vine.

Morning Glory. The first vine I ever knew was a morning glory. It grows easily as an annual from seed and will climb the tallest telephone pole and send forth lovely blue, pink, or white blossoms all summer long. It is the poor man's clematis.

TIP NUMBER NINE: *Go Native*

Now, I am not suggesting you take off your clothes and run around in your back-yard like a crazy person. Although sometimes you might feel like that after certain days at work. No, I mean give your landscape a more natural look by using more natural-looking materials for pathways, foundations, furniture, retaining walls, arbors, and trellises.

You've already started your landscape on a natural look by planting bushes, trees, and flowers that attract birds and butterflies; you wouldn't want to jar that look by putting in a concrete walkway.

Since you are going to attract nature to your backyard, you want your backyard to look a little bit more like nature, a kind of organized disorganization. You will notice that in nature, there are no straight lines; everything is curved and dangling and uneven. I am not saying your yard should look like a mess or a jungle or the aftermath of a hurricane, although at Bird Haven in Charlotte, North Carolina, several trees that were downed by a hurricane were left untouched as a perfect habitat for the hundreds of birds that live there.

Let your imagination and your good taste be your guides, but here are some of the landscape items I suggest you consider:

Natural Pathways

Instead of a concrete or asphalt sidewalk, lay down a pathway made out of wood chips. The driveway and pathways around our house are made from wood chips we get delivered by the tree crew of our local utility. They have to be replenished each year, but we are recycling matter that would otherwise have to go to the landfill, and a wood-chip driveway makes our house cooler during the heat of the

Try lining your landscape pathways with wood chips recycled from utility company tree-trimming crews, from landscape contractors, or even from your town landfill if they shred trees there. The chips have to be replenished every year or two, but they give your landscape a natural look and are a lot cheaper than slate, bricks, or other pathway materials. Try building a twig arbor out of trees you find in the woods near your home. It's a fun family activity that will add enormous beauty to your yard when the arbor is covered with blooming roses or climbing hydrangea or wisteria.

summer because it does not store heat the way concrete does. Hundreds of parks and nature centers all over the country use wood chips as pathways.

Native stone is another good alternative to concrete or store-bought paving stones. It means you have to rummage around in the woods or an abandoned quarry to find large flat stones, but that can be a good project to involve the kids. If you can't find stone, you might consider locating old paving bricks, especially those with the brickyard's name on them, to make a pathway. The streets of Springfield, Illinois, were paved with brick when I was growing up, and I still remember how much I liked their look. For both stone and brick pathways, plant fragrant low-growing herbs like thyme or chamomile in between the stones or bricks, instead of spraying the pathway with herbicides to keep the weeds down. You don't want weedy-looking grass in there, but fragrant herbs give off a nice aroma when you walk over them.

Grass also makes an attractive pathway, especially when you flip-flop your landscape and make grass the pathway between huge plantings of flowers and shrubs, rather than a huge grass lawn surrounded by flowers or hedges.

Whatever kind of pathway you choose, line it on both sides with flowers and shrubs.

Retaining Walls

Creosote-soaked railroad ties and concrete blocks are two of the retaining wall materials I hope you will not use. Creosote is a toxic chemical that can migrate out of the wood and into the soil, so besides the fact that it is unsightly, it can be dangerous. Concrete blocks are just too austere for a garden, although I am sure there are artists who can make concrete blocks look great.

I suggest building retaining walls out of natural stone that you have found in your area, or ordinary wood from the lumberyard. Yes, the wood will rot and deteriorate over time, probably several years, but it will deteriorate naturally and become compost for the soil, and it will look natural like forest wood as it gets older and older.

Decks and Furniture

I am not a big fan of decks, and I am very particular about garden furniture. I find decks to be intrusive. They seem to just jut out into the landscape like a dangling participle, and they always seem to look the same no matter how they are configured. Decks are for people who don't like to get too close to nature. If you have a deck, be sure to landscape around it. Grow a climbing rose or some other vine along the railings to soften up those hard edges. Install window boxes on the railings and fill them with herbs or exuberant nasturtiums. Think of installing a canopy over the deck either of heavy canvas or with grapevines growing up and over a wooden trellis to provide shade and quality living. I much prefer patios floored with native stone, old bricks, natural grass, or another surface to the slightly green pressure-treated wood of a deck.

Furniture should also blend in with your now more natural-looking landscape. What that means depends entirely on you. I prefer an eclectic collection of wooden Adirondack chairs or shell-back metal chairs from the 1940s to the gaily painted plastic chairs of the 1990s. But I have seen some very tasteful plastic chairs in a deep green color that look great, and, of course, those English-style garden benches are absolutely beautiful. Just think of your deck and lawn furniture as part of the landscape and not something you toss willy-nilly into the mix.

Arbors, Gates, and Posts

Arbors, gates, and posts, which if you think about it are recreations of standing, bent, and fallen trees and their overgrown vines and brush one finds in the forest, will be one area where you can really get imaginative and environmentally harmonious. I've always been a big fan of "found art," which is sculpture made out of debris found by the artist. This is really a form of recycling old discarded items. As a backyard landscaper, you can decorate your yard with found art in the form of gates, arbors, and posts from yard sales or "urban archeology" stores of items collected by dealers from old buildings. Let me give you an example.

I recently visited St. Mary's Episcopal Church in Barnstable, Massachusetts, which offers a very good example of the type of landscaping I'm talking about. One of the main gates to its public garden is flanked by two old weathered posts that were apparently scavenged out of an old fence or perhaps porch. It gives the place a worn and inviting feel that blends in perfectly with the vines and bushes that complete the entrance.

Another example was provided by my wife, who took our son on a scavenging trip into the woods behind our Hollowville home and returned with four narrow tree trunks that she is going to use to build the arbor gate that not only will invite people into our backyard but will also create a natural privacy barrier. (I get a lot of my landscaping ideas from my wife, although I usually end up doing most of the heavy lifting.)

This whole concept of using recycled materials found at yard sales, estate sales, auctions, or specialty shops, or in the woods behind your house or some other neat place, can extend to furniture, garden benches, patio planters, window boxes, and other backyard items. I know a New York painter who has assembled a collection of forty or fifty terra-cotta pots in varying sizes that he has put together in artful disarray to create a beautiful backyard accent.

TIP NUMBER TEN: *Let the Good Bugs Beat the Bad Bugs*

Luckily for us, flowers, vines, shrubs, and ornamental trees are not as prone to disease and insect attack as fruits and vegetables are. Aphids, powdery mildew, and slugs are all pretty easy to control with nontoxic or low-toxic solutions.

In fact, the best way to control those disease problems is to avoid them altogether by planting disease-resistant varieties and keeping the landscape clean and tidy. The best way to control insects is to ring the dinner bell and let their natural predators come feast on the aphids, mites, and other bugs that are causing you problems. There are even predator microorganisms that will help control diseases.

The very first thing you have to do is stop using chemicals, fungicides, herbicides, and insecticides in the garden. One of their biggest drawbacks is that they tend to destroy all the predatory bugs, birds, and fungi right along with the aphids and mildew they are meant to destroy. It is probably a good idea to stop or limit the use of botanical pesticides such as pyrethrum or rotenone as well, because they, too, can kill beneficial insects, although not to the extent that synthetic-chemical pesticides do.

More and more nontoxic and low-toxic pest control products are becoming available at local lawn-and-garden centers. If you must use insecticides, choose these. They're just as effective as their chemical counterparts and are well worth your interest.

The next step is to create a garden and landscape environment that is so wonderful the beneficial insects like ladybugs, green lacewings, predatory mites, spined soldier bugs, and others will be eager to live there. How? Stop using chemicals, add organic matter to the soil, use natural organic fertilizers, cover the ground with natural mulch, water slowly with a soaker hose, and do all the other things I have mentioned in this book. Remember in the movie *Field of Dreams* when the voice told Kevin Costner "If you build it, I will come"? That's all you have to do to attract beneficial insects to your landscape. Now, of course, you can buy beneficial insects from garden centers and several catalogues. But those bugs will not stay in your garden if you are still following a chemical program.

Finally, contact your local county Cooperative Extension office for the latest news on predatory and parasitic insects. Many universities are spending a lot of time and money studying insects to find new ways to let the good bugs beat out the bad bugs.

Here is a rogues' gallery of insects and diseases you are likely to encounter in your landscape and some of the safest ways to control them.

Insects

Aphids. One-eighth-inch-long, pear-shaped, multicolored, soft-bodied bugs that suck juices out of your plants. Control with water spray, insecticidal soap, Neem, ladybugs, spiders, and true bugs.

Japanese Beetles. Black and green ¹/2-inch-long bugs that arrive in midsummer in swarms. Pay the children a penny a piece to pick them off plants by hand and place in soapy water. You can also spray them with botanical insecticides such as Neem or pyrethrum. Pheromone-scented beetle traps do collect a lot of beetles, but it is not yet determined whether they capture your yard's beetles or the entire neighborhood's beetles. You might also try to control beetles in their grub stage underground by spreading beneficial nematodes or milky spore on the lawn.

Slugs. Control with pheromone-baited diatomaceous earth powder such as Chemfree's Insectigone, slug traps, or saucers of beer, into which the slugs do crawl and then drown. You can also sprinkle the slugs with table salt, which kills them.

Spider Mites. Almost microscopic little bugs that appear as tiny red spots on the underside of leaves. Control with insecticidal soap spray or predatory spider mites, ladybugs, or lacewings.

Thrips. Almost microscopic bugs that suck juice out of your plants and leave behind tiny black spots. Control with insecticidal soap, the botanical rotenone, ladybugs, or diatomaceous earth powder.

Diseases

Black Spot on Roses. The best way to control black spot on roses is to plant roses that are naturally resistant to black spot disease. Many of the newer varieties of hybrid tea roses tend to be more susceptible to black spot and other diseases than the older rugosa-type roses, but this is not always true. Contact your local garden center, county Cooperative Extension agent or university horticultural department and ask about rose varieties that are suitable for your climate and region and that are also disease resistant.

Other ways to control black spot are to prevent the disease from spreading up to the leaves by gathering up any fallen leaves, keeping the weeds pulled out, pruning out any dead material, and mulching heavily in winter and spring with a good dose of organic matter, preferably compost.

You might also spray the plants with a 5 percent baking soda and water solution or a sulfur-based garden fungicide.

Botrytis and Powdery Mildew. These two diseases look a lot alike. They cover the stems and leaves of infected plants with a powdery white or gray mold that makes your plants look pretty ugly. Make sure there is plenty of space between plants to allow for good air circulation, keep the beds clean and weeded, and water evenly with a soaker hose to prevent the diseases. You can plant disease-resistant varieties of several flowers if the ones you are growing continue to be plagued. You can also spray the plants with a sulfur-based garden fungicide.

Critters

The only fool-proof way to control critters such as deer, squirrels, rabbits, moles, ground hogs, and raccoons is sturdy fencing that probably should be electrified. This is a lot easier to do for a vegetable garden, which is usually grown in a square or rectangle, the perfect design for a fence. But for landscaping it is much more difficult. An electric fence in the city or suburbs can annoy your neighbors and their kids especially when they get a little shock while playing ball next door.

Trapping and other homemade remedies like soap or human hair in the landscape do not work for very long. It is difficult to hunt squirrels and woodchucks in populated areas. Probably the most effective control for them is critters of your own, namely dogs and cats.

~

One of the most important things to remember in this ten-point program is to get out there and enjoy your yard, garden, and landscape. It was a great place to be before you started this whole new program, and now it is even better. To help you enjoy your newfound space, here are some activities.

Entertain more. Set up a table and chairs and dine al fresco. Even if you only serve cocktails and appetizers before the main meal, your guests will enjoy being outdoors, and you will, too.

Buy a bird book. Since you are attracting birds, now is a good time to learn what birds are coming home to roost. There are several good bird books out there, and all will help you identify your favorites.

Help your kids build a fort. Even if you don't actually build a permanent treehouse or something elaborate like that, you can put up a make-believe fort with blankets and chairs. Even though this isn't gardening, it gets you out in the yard, it makes you study your landscape to find a good place to put the fort, and you might even do a little quick landscaping around the fort while you are there. You might like the fort and decide to plant a few flowers around it to spruce it up.

Harvest your landscape. Pick the flowers as they come into bloom and make bouquets with them. Pick the herbs and dry them for cooking or for making potpourri.

Go window shopping at garden centers. You don't have to buy anything, but nurseries and garden centers are great places for browsing, getting to know all sorts of different plants, talking with the friendly staff, and probably making new friends while you are there.

Resources

Your local lawn-and-garden center is an excellent place to shop for lawn and garden seeds, transplants, fertilizers, garden tools, power equipment, compost bins, natural pesticides, and other garden needs.

Grass Seed

A fully stocked garden center will usually have a good variety of grass seed mixtures for sale. If you are lucky, they will also have someone with the expertise and experience to know what grows best for your area.

Here is a list of several of the better-known and widely available grass seed companies and how to reach their home office if you can't find their products locally.

Agriturf, 59 Dwight Street, Hatfield, MA 01038, (413) 247–5687. Its Safe Lawn lawn mix is also recommended by the Ecologial Landscaping Association.

Gardens Alive! 5100 Schenley Place, Lawrenceburg, IN 47025, (812) 537–8650. Great selection of environmental grass seed, beneficial nematodes, fertilizers, and other garden products.

Harmony Farm Supply, 4050 Ross Road, Sebastopol, CA 95472, (707) 823–9125. Sells several mixtures, most adapted for California: California Green, with perennial rye and Kentucky bluegrass; Shade and Sun, with perennial ryegrass and two types of fescue.

Jonathan Green, P.O. Box 326, Farmingdale, NJ 07727, (800) 526–2303. A one-hundred-year-old company with a variety of lawn seed mixtures, most of them pest and disease resistant.

Lofts Seed, Inc., Bound Brook, NJ 08805, (800) 526–3890. One of the largest producers of disease- and pest-resistant grass seeds, wildflower mixtures, and other grass seeds.

Nichols Garden Nursery, 1190 North Pacific Highway, Albany, OR 97321, (503) 928–9280. Nichols offers a lawn blend it calls Ecology Lawn Mix. There are three: Northern, Dryland, and Southland. Instead of creating a lawn of strictly grass seeds, the ecology mixes add strawberry and white Dutch clovers, wild English daisies, California poppies, and other seeds to perennial ryegrasses and fine fescues.

North Country Organics, RR #1, Box 2232, Bradford, VT 05033, (802) 222–4277. Merner's Mix, its blended environmental grass seeds, one mix for northern lawns and one for southern lawns, is recommended by the Ecological Landscaping Association.

O.M. Scott & Co., 14111 Scottslawn Road, Marysville, OH 43041, (513) 644–0011 or (800) 874–7336. Scotts has a reputation for good-quality seeds, and now it is offering some of the pest- and disease-resistant types.

Pennington, P.O. Box 290, Madison, GA 30650, (800) 277–1412. The big company in the South. Look for its new Enviro blend of turf-type tall fescue.

Ringer, 9959 Valley View Road, Eden Prairie, MN 55344, (800) 654–1047. Sells four mixtures: Northern Sun Mixture, with Kentucky bluegrass and perennial ryegrass; Northern Shade Mixture, with fine fescue and shade-tolerant bluegrass; Southern Sun Blend, with a variety of Bermuda grasses; and Southern Shade Blend with three types of tall fescue.

SEED CATALOGUES

For a complete listing of most available seed catalogues, contact the National Gardening Association, 180 Flynn Avenue, Burlington, VT 05401. Ask for the *Directory of Seed & Nursery Catalogues.*

Organic and Untreated Seeds

Bountiful Gardens/Ecology Action, 5798 Ridgewood Road, Willits, CA 95490. Organic, untreated seeds and other garden items.

Heirloom Seeds

Plants of the Southwest, 1812 Second Street, Santa Fe, NM 87501, (505) 983–1548. Seeds for grass, drought-resistant wildflowers, fruits, and vegetables, many collected from Indians in the desert and mountains of New Mexico.

Johnny's Selected Seeds, Foss Hill Road, Albion, ME 04910, (207) 437–4301. The largest and most complete selection of untreated seeds.

The Natural Gardening Company, 217 San Anselmo Avenue, San Anselmo, CA 94960, (415) 456–5060. Features a large selection of gourmet vegetable seedlings grown on an organic farm near Santa Cruz, California. Includes cucumbers, alpine strawberries, melons, squash, broccoli, herbs, and tomatoes. Also offers organic seed potatoes and drought-resistant wildflowers.

Nichols Garden Nursery, 1190 North Pacific Highway, Albany, OR 97321, (503) 928–9280. Extensive selection of untreated seeds as well as some seedlings, herbs, and transplants grown under organic conditions.

Peace Seeds, 2385 Southeast Thompson Street, Corvallis, OR 97333. Very large informational catalogue.

Seed Savers Exchange, 203A Rural Avenue, Decorah, IA 52101. Not-for-profit group working to save heirloom and endangered seeds from extinction.

Southern Exposure Seed Exchange, P.O. Box 158, North Garden, VA 22959. Old-fashioned seeds producing great-tasting fruits and vegetables, many collected from descendants of nineteenth-century American immigrants. Includes the Albermarle Pippin and Esopus Spitzenburg apple trees, which were grown by Thomas Jefferson.

Lawn Mowers and Power Equipment

Most of these manufacturers have dealers near you. Some of the power equipment companies also make chippers and shredders.

This is a very small list of power tool manufacturers. For more names and information, contact the Outdoor Power Equipment Institute, 341 South Patrick Street, Alexandria, VA 22314.

American Lawn Mower Company, P.O. Box 369, Shelbyville, IN 46176, (317) 392–3615. This is the manufacturer of eight models of lightweight reel-type push mowers.

Black & Decker, 701 East Joppa Road, Towson, MD 21286, (410) 716–3900. Maker of corded and cordless mulching, bagging, and convertible mowers.

John Deere, 4401 Bland Road, Suite 200, Raleigh, NC 27609, (919) 954–6420. Maker of the Tricycler mower, which mulches and has a bagging attachment for leaf harvesting. Deere & Co. has also introduced a backyard chipper-shredder.

Ryobi America Corp., 5201 Pearman Dairy Road, Suite 1, Anderson, SC 29625-8690, (800) 525–2579. Maker of the cordless electric Mulchinator lawn mower.

Simplicity Manufacturing Co., 500 North Spring Street, Port Washington, WI 53074-5450, (414) 284–8706. Maker of three sizes of backyard chipper-shredders and two models of mulching riding mowers.

The Toro Co., 8111 Lyndale Avenue South, Minneapolis, MN 55420, (612) 888–8801. Maker of the Recycler mower, which mulches and has a bagging attachment.

Troy-Bilt/Garden Way Manufacturing, 102nd Street & 9th Avenue, Troy, NY 12180, (800) 833–6990. Offers a complete selection of environmentally oriented power equipment including mulching mowers, rear-tined garden tillers, power shredders, composters, and garden carts.

Landscape Tools and Hoses

Anchor Swan Corp., 8929 Columbus Pike, Worthington, OH 43085, (800) 848–8707. Manufacturer of Earthquencher soaker hose and other high-quality hoses.

Aquapore Moisture Systems, 610 South 80th Avenue, Phoenix, AZ 85043, (800) 426–8419. Maker of the Moisture Master soaker hose and other quality hoses.

Corona Clipper Co., 1540 East Sixth Stret, P.O. Box 1388, Corona CA 91718-1388, (714) 737–6515. Good selection of high-quality pruning shears, saws, loppers, and other hand-held landscape tools.

Denman & Company, 2923 Saturn Street, Brea, CA 92621, (714) 524–0668. Offers a selected group of tools as well as gardening apparel.

The Great American Rain Barrel Co., 295 Maverick Street, East Boston, MA 02128, (617) 569–3690. Maker of rain barrels.

Harmony Farm Supply, P.O. Box 460, Graton, CA 95444, (707) 823–9125. Offers an extensive selection of irrigation equipment and a comprehensive collection of hand and power tools.

Kinsman Company, River Road, Point Pleasant, PA 18950, (215) 297–5613. Has a very good selection of English- and American-made shovels, rakes, hand tools, and watering cans.

A. M. Leonard, Inc., 6665 Spiker Road, Piqua, OH 45356, (800) 543–8955. Publishes a mail order catalogue for the landscape industry. Good tools at good prices.

Smith & Hawken, 25 Corte Madera, Mill Valley, CA 94941, (415) 383–2000. Has classy garden tools that are also very sturdy and practical. In addition to hand-crafted shovels and spades, some esoterica such as a poacher's spade, a woodman's pal, a brush slasher, and a Japanese hatchet.

True Temper, 465 Railroad Avenue, P.O. Box 8859, Camp Hill, PA 17001–8859, (717) 737–1500. Maker of full line of pruners, shovels, rakes and other high-quality hand tools for lawn care and landscaping.

Walter Nicke Co., P.O. Box 433, Topsfield, MA 01983, (508) 887–3388. Catalogue lists over three hundred tools including some great hand pruning shears and clippers.

Pest Control

Bonide Products, Inc., 2 Wurz Avenue, Yorkville, NY 13495. Offers a wide variety of biological and least toxic pest controls at lawn-and-garden centers. Bonide also makes synthetic chemical pesticides, so be sure to read the labels carefully.

Safer/Ringer, 9959 Valley View Road, Eden Prairie, MN 55344. Safer, now owned by Ringer, makes a wide variety of biological controls such as Neem and Bt, as well as many least toxic pest controls that are commonly available at lawn-and-garden centers and through catalogues, including their own. For a free catalogue, call (800) 654–1047.

The following catalogues offer extensive collections of biological and least toxic pest controls, as well as traps, beneficial insects, barriers, row covers, birdhouses, and other tools to help you control unwanted garden pests. Some of these catalogues can serve as textbooks on the subject.

Gardens Alive! Natural Gardening Research Center, 5100 Schenley Place, Lawrenceburg, IN 47025, (812) 537–8650. These people are doing a lot of their own research on the effectiveness of natural pest controls. Especially good for midwestern gardeners.

Harmony Farm Supply, P.O. Box 460, Graton, CA 95444, (707) 823–9125. Especially deep catalogue with nice selection of insect traps.

Integrated Fertility Management, 333 Ohme Gardens Road, Wenatchee, WA 98801, (509) 662–3179. Especially good for fruit trees.

The Natural Gardening Company, 217 San Anselmo Avenue, San Anselmo, CA 94960, (415) 456–5060. Limited but well-chosen selection of natural pest controls suitable for small family gardens.

An increasing number of seed catalogues are beginning to carry or feature natural pest control products. Three favorites are:

W. Atlee Burpee & Co., 300 Park Avenue, Warminster, PA 18974, (215) 674–9633 or (800) 888–1447. Good selection of natural controls including beneficial insects.

Johnny's Selected Seeds, Foss Hill Road, Albion, ME 04910–9731, (207) 437–4301. Information about the various biological and botanical insecticides is very informative.

J.W. Jung Seed Co., Randolph, WI 53957, (414) 326–4100. A family business with good prices since 1907. Good selection of natural pest controls as well as live traps and sticky strips.

Gardening and Environmental Associations

The American Forestry Association, P.O. Box 2000, Washington, DC 20036, (202) 667–3300.

Clean Water Action Project, 317 Pennsylvania Avenue SE, Washington, DC 20003, (202) 547–1196.

Environmental Action, 1525 New Hampshire Avenue NW, Washington, DC 20036, (202) 745–4870.

Environmental Defense Fund, 257 Park Avenue South, New York, NY 10010, (212) 505–2100.

Garden Clubs of America, 598 Madison Avenue, New York, NY 10022, (212) 753–8287.

National Arbor Day Foundation, 100 Arbor Drive, Nebraska City, NE 68410, (402) 474–5655.

National Audubon Society, 950 Third Avenue, New York, NY 10022, (212) 832–3200.

National Coalition Against the Misuse of Pesticides, 530 7th Street SE, Washington, DC 20003, (202) 543–5450.

National Gardening Association, 180 Flynn Avenue, Burlington, VT 05401, (802) 863–1308.

National Solid Wastes Management Association, 1730 Rhode Island Avenue NW, Washington, DC. 20036, (202) 659–4613.

National Wildlife Federation, 1400 16th Street SW, Washington, DC 20036, (202) 797–6800.

Natural Resources Defense Council, 122 East 42nd Street, New York, NY 10168, (212) 949–0049.

Index

H

Herbicides, to kill grass/weeds, 54
Holly, 159
Hoses, 38
    nozzles, 38
    regular hose, 38
    soaker hose, 38, 143–44
    sources for, 169
Hosta, 157
Hummingbirds, 147
Humus Builder, 73–74

I

Insects, 22–24
    beneficial nematodes, 85, 86, 87, 91
    beneficial types, 148, 162–63
    biological pesticides, 23–24
    endophytes as repellent, 18
    preventive measures, 23, 114, 162
    types of harmful insects, 163–64
    See also specific insect pests
Iris, 157
Irrigation system, 19

J

Japanese beetles, 23–24

K

Kentucky bluegrass, 16, 17, 18, 47, 52, 56, 98, 99, 101

L

Landscape
    assessment of ecology of, 134–35
    edible landscaping, 138–39

    for energy conservation, 151–54
    flowers, 155–58
    ground covers, 136–38
    natural landscape, 159–62
    organic matter, adding to, 141–42
    pH for landscape plants, 141
    shrubs, 158
    trees, 149–51
    vines, 159
    watering, 143–45
    wildflowers, 138
    wildlife in, 145–49
    xeriscaping, 145
Lawn mowers
    electric mowers, 26
    gas-powered mulching mowers, 26–28
    maintenance of power mowers, 29
    manufacturers of, 168–69
    reel mowers, 26
    riding mowers, 26
    safety guidelines for power mowers, 28
Lawn roller, 52
Lawns
    basic care of, 139–40
    guidelines for assessment of, 45–46
    rehabilitation of, 51–54
    remodeling of, 54–60
    time for planting, 18, 50–51
    types of grasses, 46–50
Leaves, source of organic matter, 8–9, 43–44
Lilac, 159
Lily, 157
Lily of the valley, 157
Limestone, to correct soil pH balance, 6–7, 50, 141
Lopping shears, 34–35

M

Manure
  and composting, 65
  disadvantages of, 56
Marigold, 158
Midwest. *See* Northeast/Midwest/Pacific
  Northwest
Mock orange, 159
Mole crickets, 101, 102
Morning glory, 159
Mountain laurel, 159
Mowing grass
  first mowing, 53
  See also Lawn mowers
Mulch
  recommended type, 57
  for water conservation, 144

N

Nasturtium, 159
Natural landscape, 159–62
  arbors/gates/posts, 161–62
  pathways, 159–60
  retaining walls, 161
Neem, 23
Nematodes, 85, 86, 87, 91, 98, 102, 115,
  127
Northeast/Midwest/Pacific Northwest
  April activities, 87
  August activities, 91
  climate of, 83
  December activities, 94
  February activities, 85
  grasses for, 83
  January activities, 84
  July activities, 90

June activities, 89
March activities, 86
May activities, 88
November activities, 94
October activities, 93
September activities, 92
soil of, 83
Nozzles, for hoses, 38

O

Organic fertilizers, 10–14, 142–43
  advantages of, 12–13
  benefits of, 142–43
  brand names of, 12
  cost factors, 13
  grass-starter, 52
  sources for, 14
  types of ingredients in, 12, 13
Organic matter, 8–9
  adding to garden, 141
  grass clippings, 8–9, 15
  importance of, 141
  leaves, 8–9
  top-dressing method, 9
  types of, 141
  See also Organic fertilizers
Overseeding, 51–52
  process of, 51–52
  quantity of extra seed in, 52

P

Pacific Northwest. *See*
  Northeast/Midwest/Pacific Northwest
Pathways, 159–60
  materials for, 160
Peppers, in edible landscape, 139

About the Author

LAURENCE SOMBKE, "The Environmental Gardener," is a newspaper and magazine columnist, the host of the home video *Beautiful Easy Lawns,* and author of six books on gardening, food, and the environment, among them *Beautiful Easy Gardens* (Globe Pequot) and *The Environmental Gardener* (MasterMedia). In addition to his weekly segment on "The Environment Show," a nationally syndicated public radio program, he is also a frequent speaker at botanical gardens, horticultural societies, and garden and flower shows. He has appeared on "The Today Show" as well as on dozens of other radio and TV programs in the United States and Canada and has been an environmental gardening consultant to many companies. Larry lives and gardens with his family in Claverack, Columbia County, New York.

Gardening

From lush picture books to no-nonsense practical manuals, here is a variety of beautifully produced titles on many aspects of gardening. Each of the gardening books listed is by an expert in his or her field and will provide hours of gardening enjoyment for expert and novice gardeners. Please check your local bookstore for other fine Globe Pequot Press titles, which include:

Beautiful Easy Gardens, $24.95, paper, $15.95

The Victory Garden Kids' Book, $15.95

Landscaping That Saves Energy and Dollars, $16.95

Efficient Vegetable Gardening, $14.95

The Naturalist's Garden, $15.95

The Wildflower Meadow Book, $16.95

The National Trust Book of Wild Flower Gardening, $25.95

Garden Flower Folklore, $19.95

Wildflower Folklore, $23.95; paper, $14.95

Folklore of Trees and Shrubs, $24.95

Herbs, $19.95

Dahlias, $19.95

Rhododendrons, $19.95

Fuchsias, $19.95

Climbing Roses, $19.95

Modern Garden Roses, $19.95

Azaleas, $19.95

Auriculas, $19.95

Magnolias, $19.95

The Movable Garden, $15.95

Garden Smarts, $12.95

Simple Garden Projects, $19.95

Windowbox Gardening, $22.95

Perennial Gardens, $17.95

To order any of these titles with MasterCard or Visa, call toll-free 1-800-243-0495; in Connecticut call 1-800-962-0973. Free shipping for orders of three or more books. Shipping charge of $3.00 per book for one or two books ordered. Connecticut residents add sales tax. Ask for your free catalogue of Globe Pequot's quality books on recreation, travel, nature, gardening, cooking, crafts, and more. Prices and availability subject to change.